REPRESENTATION AND PROCESSING OF SPATIAL EXPRESSIONS

REPRESENTATION AND PROCESSING OF SPATIAL EXPRESSIONS

Edited by

Patrick Olivier
University of Wales, Aberystwyth

Klaus-Peter Gapp
Universität des Saarlandes,
Saarbrücken, Germany

LEA LAWRENCE ERLBAUM ASSOCIATES, PUBLISHERS
1998 Mahwah, New Jersey   London

Lawrence Erlbaum Associates, Inc., Publishers
10 Industrial Avenue
Mahwah, New Jersey 07430

Cover design by Kathryn Houghtaling Lacey

Library of Congress Cataloging-in-Publication Data

Representation and processing of spatial expressions / edited by
    Patrick Olivier and Klaus-Peter Gapp.
        p.   cm.
    Includes bibliographical references and index.
    ISBN 0-8058-2285-2 (alk. paper).
    1. Artificial intelligence. 2. Cognition. 3. Spatial analysis.
I. Olivier, Patrick. II. Gapp, Klaus-Peter.
Q335.R48 1997
006.3—dc21   97-4129
                                                          CIP

Books published by Lawrence Erlbaum Associates are printed on acid-free paper,
and their bindings are chosen for strength and durability.

Printed in the United States of America
10 9 8 7 6 5 4 3 2 1

Contents

Preface

Coping with spatial expressions in a cognitively plausible way is a crucial problem in a number of research fields, in particular, cognitive science, artificial intelligence, psychology, and linguistics. This volume contains both a set of theoretical analyses and accounts of applications that deal with the problems of representing and processing of spatial expressions, including: dialogue systems for understanding the use of mental images; interfaces to CAD (computer-assisted design) and multimedia systems, for example, natural language querying of photographic databases; speech-driven design and assembly; machine translation systems; spatial queries for geographic information systems; and systems that generate spatial descriptions on the basis of maps, cognitive maps, or other spatial representations, for example, intelligent vehicle navigation systems.

Though there have been many different approaches to the representation and processing of spatial expressions, most existing computational characterizations have so far been restricted to particularly narrow problem domains, usually specific spatial contexts determined by overall system goals. To date, artificial intelligence research in this field has rarely taken advantage of studies of language and spatial cognition carried out by the cognitive science community. One of the fundamental aims of this book is to bring together research from both disciplines in the belief that artificial intelligence has much to gain from an appreciation of cognitive theories.

This volume primarily consists of papers accepted for the Fourteenth International Joint Conference on Artificial Intelligence Workshop on the Representation and Processing of Spatial Expressions that took place in

Montreal, Canada, in August 1995. In addition, papers by the two distinguished invited speakers at the workshop, the linguist Annette Herskovits from Boston University and the psychologist Barbara Landau from the University of California, Irvine, and the University of Delaware, characterize the topic of spatial expressions from the perspective of their own particular disciplines. Two extra papers were also specifically solicited from Hilary Buxton and Richard J. Howarth of the University of Sussex, and from Amitabha Mukerjee of the Indian Institute of Technology, Kanpur. Mukerjee's paper opens this volume and is particularly significant as it is the first comprehensive review and assessment of representational schemes for spatial expressions.

We are grateful to all the people who have helped shape the topic and organize the workshop and the Fourteenth International Joint Conference on Artificial Intelligence, the members of the organizing committee, and the additional reviewers. These include Jugal Kalita, Paul McKevitt, Amitabha Mukerjee, Junichi Tsujii, Laure Vieu, Jeff Siskind, Wolfgang Wahlster, and Yorick Wilks. We should also like to offer particular thanks to our respective bosses, Wolfgang Wahlster and Frank Bott, for allowing us to commit the necessary time and resources to this project.

—*Klaus-Peter Gapp*
—*Patrick Olivier*
Saarbrücken and Aberystwyth, July 1996

Neat Versus Scruffy: A Review of Computational Models for Spatial Expressions

Amitabha Mukerjee
Indian Institute of Technology Kanpur

This chapter reviews the computational modalities for representing spatial expressions. The focus is on computational or synthetic models, as opposed to cognitive or linguistic aspects. A number of formalisms based on qualitative paradigms such as topology seek to discretize space into regions based on alignment or tangency. This approach may be termed as "neat." Yet a large class of spatial expressions (e.g., "near," "move away from") do not involve alignments, and even some that do (e.g., "in front of") have gradations. These gradations in a continuum are clearly less "neat"; we may view these as "scruffy"; there appears to be a growing trend toward the scruffy in recent times. The distinctions between the different models reviewed are demonstrated using the canonical example of "the chair in front of the desk."

1. INTRODUCTION

The philosopher Wittgenstein, when young (*Tractatus*, 1921), claimed to have provided a "final solution" to the problem of logic and language, based on a set of primitives that describe the boundaries of what can be expressed. In essence, this approach claims that thought is grounded on a set of distinct symbols, a "neat" discretization. In later years (*Philosophical Investigations*, 1953) he himself rejected this view, insisting that language can properly be considered only with respect to its use in specific behavioral contexts ("scruff"). Nuallain and Smith (1995) compared this famous transformation from neat to scruff to the change of paradigm within the AI (artificial intelligence) community from a primitive-based model of semantics to a more context sensitive one. The contradiction between the neat and the scruffy holds probably over the entire domain of artificial intelligence, and certainly to the topic of this survey—the computational modeling of spatial expressions.

Representing space has a rich history in the physical sciences, where many representations were developed for encoding orientations, points, vectors, rigid body motions, and so forth. These representations are, as a class, continuous and serve to locate objects in a quantitative framework— for example, in a multidimensional space of coordinates. Any assignment of values to the coordinates results in a unique and precise location for the entity.

Spatial expressions, on the other hand, operate on a loose partitioning of the domain, a discretization into regions such as "in front/back" or "near/far," where the location, and even the partitioning itself, is specified with a large degree of ambiguity. The task of finding suitable representations for this less precise description has turned out to be considerably more difficult than was initially realized. The emphasis has been on creating discretizations that matched the linguistic partitions. Unfortunately, each discretization that was created, for example, the many versions of the "ON" primitive in the blocks world, attempted to enforce the loose partitioning of a spatial expression onto the straitjacket of strict, numerical distinctions. Representations proliferated, more or less on an ad hoc basis. With the emergence of qualitative reasoning, the partitioning was done more systematically, mapping the complete set of distinctions deriving from a given set of landmarks, and treating all zones between the landmarks as indistinguishable.

1.1. The Evolution of the "Neat"

Very early AI created discretizations on an ad hoc basis, for example, by defining cells that were sensitive to the choice of origin and frame size. Discretizations mapped the continuous world into a set of discrete symbols, and the partitions were defined by the programmer based on pragmatic considerations of what appeared to suit the theory, as opposed to any systematic basis; such models may be called gerrymandered discretizations and remind one of the classification of the heavens in the work of Ptolemy, say. An example of this is the proliferation of interpretations that ON(A, B) had in the early Blocks World models (see Fig. 1.1).

More systematic discretizations were achieved by defining the complete set of partitioning based on landmarks. For example, in one-dimensional space Allen (1983) used a set of two landmarks (endpoints) in the

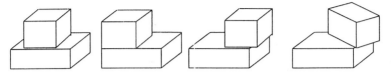

FIG. 1.1. *Where does ON(A,B) end?* This was a popular predicate in early AI and had many interpretations, depending on the flexibility of the programming. Some versions rejected overshoot (c) and most did not admit rotation (d). Partitionings were chosen to be convenient to the programmer—a gerrymandering approach common to the early stages of all sciences, forced by the belief that "neat" partitions are the basis for all meaning.

reference object (RO); the two endpoints of the located object (LO) were described completely in terms of these as a set of 13 possible orderings or arrangements (see Fig. 1.10, discussed in more detail later). Such thinking extends to spatial situations with a one-dimensional envisionment, as in the road scene of Fig. 1.2. Similarly, in higher dimensions, all possible visibility orderings based on a set of points can be modeled as an arrangement of visible points—this results in a partitioning of the plane into a set of qualitative zones: lines (two or more points aligned on same line), points (two or more lines intersecting), and regions (no two points aligned) (Kuipers, 1977; Schlieder, 1994). For example, in the four-landmark case (Fig. 1.2b), there are 23 lines, 7 points, and 18 regions based on visibility orderings on the landmarks ABCD. The partitioning in these models was based on the alignment between some fundamental entities,

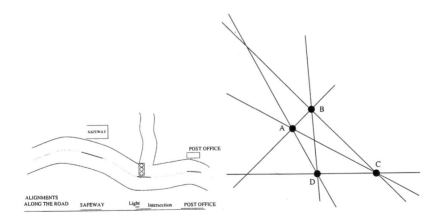

FIG. 1.2. *Alignment-based discretizations. (a) Position:* "The post office is past the Safeway store" as a linear interval model. *(b) Orientation:* A set of points partitions the domain into regions based on visibility arrangements. Multipoint alignments result in lower dimensional entities like lines and points.

such as two points in the interval model, or two orientations in the visibility-ordering model.

Qualitative reasoning models that create discretizations based on alignment now have a varied set of representational structures, many of them involving considerations of topology, focusing on connectivity rather than position (Cohn, Randell, Cui, & Bennett, 1993; Egenhofer & Franzosa, 1991). Expressions involving tangency, or transition from contact to no-contact situations, can be expressed in terms of such predicates. Other models define local frames on the objects based on intrinsic, extrinsic, or deictic (or other) considerations, which can be used to combine position and orientation (Forbus, Nielsen, & Faltings, 1991; Hernández, 1994; Mukerjee & Joe, 1990).

Within the class of alignment-based models, the systematic treatment of all possible relations resulted in much better handling of spatial expressions, especially in the zones near contact. For example, the ON relation in the orthogonal blocks world can be handled elegantly by considering a multidimensional projection model, instead of subjectively chosen predicates like "not-touches."

To simplify things tremendously, qualitative models may be thought of as multidimensional orderings of point-sets; these are in general very expressive in the region near contact or alignment—here slight changes in spatial position or orientation result in changes of order, and generate topologically relevant information. However, qualitative models are not very meaningful for noncontact or nonalignment positions, such as in the spatial prepositions like "near" or "between" or in expressions such as "moved farther away from" or "veer sharply to the left." These relations involve no changes in the alignments or orderings of the objects, and are changes of measure rather than of quality; consequently they cannot be modeled with purely qualitative formalisms. Experience reveals that many projective spatial prepositions such as "in front of" also have a gradation; for example, in Fig. 1.3, the chair A is more "in front of" the desk than the chair E, say. In a discretized model for the desk, as in the projective zones of Fig. 1.4, this distinction cannot be expressed.

1.2. From Neat to Scruffy

Although standard quantitative models are very well developed and powerful, information such as desk at (105, -33) oriented at 27.2° cannot, in itself, illuminate a spatial expression such as "in front of the desk." One of the scruff solutions here is to view the spatial relation as a fuzzy class over the quantitative space based on a *measure* defined on the continuum and not as a discrete set. An example is a possibilistic measure—one that is associated with a function such as a membership function

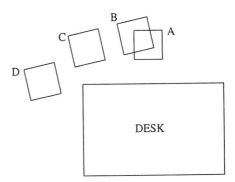

FIG. 1.3. *"The chair is in front of the desk":* The degree of in-frontness is different for positions A through D. Distance is a factor as in the difference between A and E. The orientation of the chair is also significant. Purely projective models create neat discretizations that cannot distinguish these shades of meaning within each projective zone.

used in fuzzy reasoning—or, more appropriately in spatial reasoning, as what is known as the potential field. A new class of models, inspired by work in robotics, uses potential fields to model possible locations of its objects (Gapp, 1994; Olivier, Maeda, & Tsuji, 1994; Yamada, 1993); for example, the in-front-of field can be thought of as a field with a shallow depression near the more likely positions of "in front," such as position

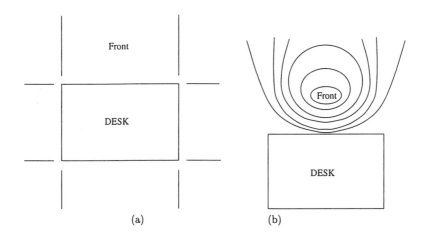

FIG. 1.4. *2-D Space discretizations for a desk:* (a) Neat: A projective discretization defines a region with sharp boundaries. (b) Scruffy: A function in a continuum model defines a degree of "in-front-ness"; lines of equal strength (potential) are shown, decreasing away from the central "sweet spot." Similar potentials are defined for other regions such as "left" and "near."

A in Fig. 1.3. A marble rolling down this bowl-shaped field would equilibrate around the more likely positions of a chair "in front" of the desk. The potential field model is defined through a set of parameters that can be tuned for different scales, distances, and orientation ambiguities. Other measures seek to preserve some aspects of the projective model by computing the degree of overlap of an object with the discretized regions (see chap. 6 of this volume). On the whole these models may be thought of as a quantitative measure operating on a fundamental qualitative distinction; together they are beginning to model a class of linguistic expression that was out of the reach of either purely qualitative or purely quantitative models.

Thus we find a transition from the earlier, strict, "neater" discretizations of space based on alignment, to the newer, looser, "scruffier" discretizations based on measures. In the new formalism, a chair at 45° is "in front" as well as "left of" in about equal measures; the partitionings overlap, as in much of the semantics of spatial expressions. Computationally, this distinction arises from a fundamental difference in the nature of the constraint used—a discrete constraint ($x < K$) results in a sharp demarcation; a fuzzy discretization ($membership(x) = f(x - K)$) results in a looser partitioning. Although the latter is probably closer to the semantics of much spatial expression, clearly one pays a computational penalty in using such complex field models.

In a deeper sense, the formalisms under which spatial expressions are understood and modeled are also undergoing change. In the 10-year-old influential work by Herskovits (1986), the mechanism proposed for handling spatial prepositions was in terms of first identifying an "ideal" meaning determined through geometric relations, and subsequent modifications by a set of "use types" to define context. Recently, Coventry has strongly challenged this position (see chap. 15 of this volume), stating that neither minimal nor maximal specifications of meaning can cover the variability in spatial expression; this can be unified only by considering the functional relation between the reference object and located object. One of his examples is illustrative: Consider the topmost apple of the heap "in" a bowl; the same apple in the same position is not considered "in" if suspended at the same position (see Fig. 15.4). Can all such variations be captured by use-type definitions? The new thinking appears to demonstrate an increasing shift away from the ideal meaning and the neat category toward a scruffier definition that is dependent on being able to satisfy a function, to a *degree* (Fig. 1.5).

This chapter is organized in four sections. First we look at a taxonomy for the entire field of qualitative spatial reasoning, which has now come to embrace both the alignment and the continuum models (section 2). This provides an overview, with quick sketches of some of the models

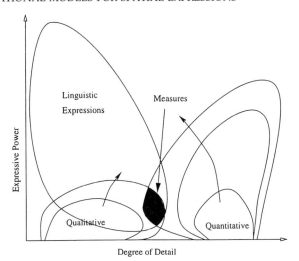

FIG. 1.5. *Measure in spatial relations:* As qualitative and quantitative models evolve, greater needs of interaction are perceived. A quantitative *measure* of qualitative attributes such as the degree of in-front-ness can be modeled as a possibilistic measure, or a potential field. This gradation, although quantitative, can be defined using a qualitatively partitioned domain. It is likely that further progress in spatial reasoning would involve more such quantitative-qualitative hybrids, which result in greater coverage of spatial expressions.

based on examples drawn from the chair/desk scenario. Sections 3 and 4 elaborate on some of the popular alignment-based discretizations (neat) and some of the newer continuum models (scruff). One key issue in modeling spatial expressions involves the shape of its objects, which is handled in section 5.1. Next we discuss the applicability of some of these models in application problems such as cognitive image generation and route description, image labeling, diagram understanding, and so forth.

2. QUALITATIVE SPATIAL REASONING: A TAXONOMY

In a review of qualitative approaches to spatial reasoning, Freksa and Röhrig (1993) classified the approaches for spatial reasoning based on the domain of the problem handled; that is, what kinds of questions can be answered by the model. To put it very roughly, they classified procedures into those dealing with (a) point positions, (b) orientations, (c) topology, and (d) position and orientation, making additional distinctions on the manner of attaching frames and so on. At that time, barely 3 years ago, all these models were essentially "neat," defined by alignment. Buisson (1989) attempted a separate classification of spatial relations based on topology, vector directions, metrics, and Euclidean angles; this contains a mix of quantitative and qualitative measures.

Today, we may look at this decomposition from another light; instead of asking the question of the approaches that have evolved, we could try to ask the question of the spatial reasoning needs; that is, what should be modeled in order to answer questions about space? Some of the things we need to model are positions and orientations, shape, scale and dimensionality of the problem, multiple objects, and so forth. This list does not claim to be complete, but merely to serve as a basis for integrating the many diverse models that have been worked on to date. Following this taxonomy, we take a detailed look at the models.

2.1. Functional Classification

1. *Dimensionality.* The space of the chair/desk can be one-, two-, or three-dimensional.
    a. Neat: Discrete one-dimensional models include interval algebra (Allen, 1983), which can be extended under various assumptions to obtain different higher dimensional alignment models, such as the point-based arrangement models (Fig. 1.6).
    b. Scruffy: Instead of sharply bounded intervals, one may associate measures on the same qualitative zones, resulting in fuzzy sets in one dimension, and potential-based continuum models in higher dimensions. Most higher dimensional models need to make some approximations on the shape, scale, or frames of the objects.
2. *Scale.* Issues relate to the size of objects, but more important, to multiple levels of abstraction (scale-space) and the connectivity between models at different scales. As the scale becomes macroscopic, the chair and desk may fuse and then disappear, and in the microdirection, they may eventually constitute different fields of view and lose their relation. See Topaloglou (1994), Kaufman (1991), and Mukerjee and Schnorrenberg (1991) for different approaches to scale models. Quadtrees are popular models of multiscale analysis. Olivier, Ormsby, and Nakata (1995) used multiple scales on a grid model.
3. *Shape.* Consider the models for physical shape.
    a. Point: Ignores spatial extent (many potential models). Point-visibility-based models or their generalization, the oriented matroid

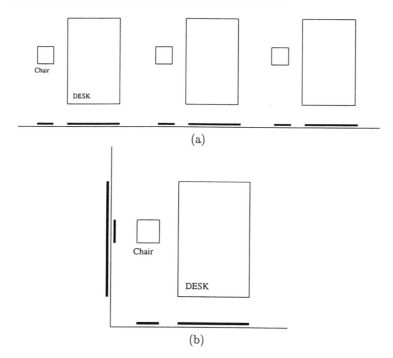

FIG. 1.6. *Dimensionality*: "Several desks in a row" can have a 1-D (linear) instantiation, but "the chair is in front of the desk" may be considered as 2-D (planar) (b), whereas "the work posture is very comfortable with this layout" involves some 3-D (volume) aspects. The projections onto two axes lead to separate interval decompositions. This type of model is useful in naturally orthogonal domains, such as document analysis (Fujihara & Mukerjee, 1991).

theory (Björner, Las Vergnas, Sturmfels, White, & Ziegler, 1993), fall into this class.

b. Rectangle: Approximations of shape by a rectangular enclosure. See section 3.2.

  * Orthogonal world: All boxes have the same global orientation (Glasgow, 1993; Rajagopalan, 1994).
  * Local front: Viewer (deictic), object (intrinsic), or external characteristics used to orient the rectangle (Jungert, 1994; Mukerjee & Joe, 1990). The assignment of frames and interchange between such frameworks is a fundamental problem in modeling spatial expressions (see chap. 14 of this volume).
  * Minimum enclosure: Use the minimal-area box.

c. Direct shape: Based on quantitative models, abstraction on these is difficult.

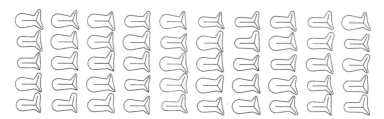

FIG. 1.7. *Axis-based shape models:* "A fish has a bulging shape in the front and a flaring tail at the back." An axis-based model can have its axes perturbed, resulting in a class of shapes (Agrawal et al., 1995).

* Boundary-based: Face/edge/vertex hierarchies, or chain-code variations—for example, using curvature versus arc-length or other intrinsic geometry concepts.
* Axis-based: Using the medial axis as a model of shape (Agrawal, Mukerjee, & Deb, 1995) (Fig. 1.7).
* Harmonic models: Uses Fourier (Forsythe, 1990) or sinusoidal space transformations (Pratt, 1989).
d. Constructive: Shape as a combination of simpler primitives.
* Glue: Simple primitives with "attachment surfaces" (Biederman, 1990; Dickinson, Pentland, & Rosenfeld, 1992).
* Boolean: Permit shape negation. Popular input interface for geometric modelers.
* Convex difference model shapes as the convex hull minus the pockets, all constrained by topological primitives like touches-in-chain, is-connected, and so on (Cohn, 1995).
4. *Topology.* Are the objects in contact? If so are they touching, overlapping, and so forth? Clementini and Di Felice (1995) provide a review of approaches using such topological considerations.
a. Nature of contact: Theoretical problems such as two closed-set objects cannot have points shared, or if open sets, then cannot have points arbitrarily close (Asher & Vieu, 1995; Hayes, 1985). Therefore, objects cannot touch.
b. Relation sets: Sets of topological relations. Grigni, Papadias, and Papadimitrou (1995) investigated the computational satisfiability for different simplifications of these topological models.
* Boundary-based: Considers the intersections of boundaries and interiors, resulting in eight relations (Egenhofer & Franzosa, 1991). Considers the nature of the nonempty.
* Connectivity-based: According to Cohn et al. (1993), the seven primitives in this model are related to seven primary relations

in interval algebra (also see Ligozat, 1994). An added aspect is reasoning based on the convex hull.

5. *Position and Orientation.* Considers the relative pose between objects.
   a. Discretized sets: Abstraction classes for spatial position are modeled as discrete sets.
      * Orthogonal frames: Uses a global x-y-z frame for relating positions of objects (Glasgow, 1993; Mukerjee & Mittal, 1995; Rajagopalan, 1994).
      * Orientation: Frames can have any orientation. Orientation is cyclically ordered. Models vary between four and eight primary directions (Fig. 1.8).
      * Topology and orientation: Topology together with orientation can be used to generate a general framework, as in Hernández (1994).
      * Position and orientation: The interaction between these makes this problem particularly difficult (Jungert, 1994; Mukerjee & Joe, 1990).
      * Array models: Positions encoded on variably sized arrays (Glasgow, 1993).
      * Oriented matroids: Oriented matroid theory (Björner et al., 1993) provides an elegant generalization for the n-dimensional alignment partitioning. The signed chirotopes result in the same distinctions as in visibility-based partitioning (Fig. 1.2b).
      * Voronoi diagram: Divides up the space into convex regions closest to each object; adjacencies contain positional information (see chap. 10 of this volume).
      * Distance: Discretize distances (the noncontact region) into sharp regions of near, very near, and so on (Hernández, Clementini, & Di Felice, 1995).
   b. Continuum measures: Classes defined in terms of measures, with fuzzy boundaries, and possible overlap.

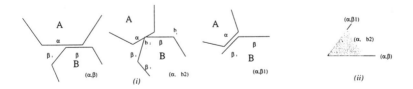

FIG. 1.8. *Orientation as interval algebra:* Orientation changes of one object with respect to another can be modeled as interval relations between angular intervals. (Unlike Fig. 1.10, these are cyclic; i.e., the two ends [before and after] are the same.)

* Potential models: Relations such as "in front of" holds with higher strength some distance from the front of the chair and falls away on all sides. This notion can be captured as a two-dimensional (2-D) field, which may be thought of as a possibilistic measure or a 2-D fuzzy function (Gapp, 1994; Yamada, Yamamoto, Ikeda, Nishida, & Doshita, 1992) (Fig. 1.9).

* Weighted sum over discretization: Here one uses the overlaps of an object with a discrete set to obtain the continuum measure (see chaps. 6 and 10 of this volume).

   c. Constraint propagation: Given constraints on positions, how to generate feasible placements for the objects. This has a very large literature of its own (Bhansali & Kramer, 1994; Charman, 1994; Olivier et al., 1994).

   d. Nonalignment positions: Such as "between," "near," and so forth; these are primarily modeled using continuum measures (Gapp, 1994).

6. *Multiple Objects.*

   a. Interaction: Multiple objects may interact indirectly: "He turned the corner" may be considered in an orthogonal (lattice) framework (Narayanan, Suwa, & Motoda, 1994; Olivier et al., 1995) or by focusing on the interaction between shapes in general (Mukerjee, Agarwal, & Bhatia, 1995).

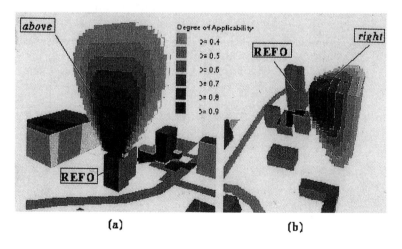

        (a)   (b)

FIG. 1.9. *Continuum measures based on potential:* Here the above and right relations, normally considered as discrete (projective), are modeled using continuum measures, which discriminate the "degree" of above, and so on. One constraint in most potential models is that the LO has to be treated as a point. From Gapp (1994). Reprinted by permission.

   b. Patterns: Repetitive patterns and arrangements; "place the mar-
      bles on a circle."
   c. Kinematics: Motion constraints between objects; configuration
      space models (Forbus et al., 1991; Olivier et al., 1995; Sacks &
      Joskowicz, 1993).
7. *Integrating Time.* Dynamic scenes also define locations; for example,
   "He went around the desk" not only defines the motion, but also
   the desk position (e.g., it cannot be flush at a corner). Due to poor
   modeling of kinematics, this type of model is yet to emerge. Some
   of the diagram-based work models simple motions as in motion
   down a plane (Rajagopalan, 1994).

Most models span several of these categories—these are more like the
dimensionalities of the problem, but we have attempted to use this taxonomy to organize the structure of this review.

The preceding taxonomy contains a large omission. When we talk of
spatial reasoning in this chapter and otherwise in the context of AI or spatial
expressions, we often ignore the large body of literature in geometry, vector
algebra, pattern recognition, solid modeling, graphics, and other fields that
also claim to address much the same issues. I see the prime difference
between these and the models that are emerging in AI as that of a capability
for abstraction, although even here, the goals of multiscale paradigms of
image processing seem to be extremely similar. Models that eschew all
forms of quantitative measures are often talked of as "qualitative" though
this is by no means a clearly defined boundary. In any case, many of the
models of spatial location—for example, the vast majority of which deal
with noncontact situations—involve metrics that are captured in the potential models, which are clearly an offshoot of the quantitative paradigm. The
dichotomy in vocabulary and concepts that has emerged between spatial
expression modeling and the traditional (physical sciences) models of space
is undesirable and something that we should strive to reduce.

In addition to the aforementioned taxonomy, Raman Rajagopalan
(1994) pointed out that another criterion for comparing computational
models of space is the assumed level of incompleteness in the input. For
example, in the absence of specific locations, neither the potentials nor
the configuration space models are useful. Clearly, this is an important
factor, and will result in a different framework of comparison. Another
distinction not considered in this review is that of mixing qualitative and
quantitative models—the term "hybrid models" can mean carrying both
qualitative and quantitative data simultaneously (Forbus et al., 1991;
Kautz & Ladkin, 1991), or it may mean carrying data at various levels of
granularity (Mukerjee & Mittal, 1995).

3. DISCRETIZING SPACE: ALIGNMENT MODELS

Spatial reasoning, reduced to one dimension (Fig. 1.10), can be modeled
using the set of 13 interval relations in interval logic (Allen, 1983). Point-
based versions of this logic are also well known (Guesgen, 1989; Kumar
& Mukerjee, 1987; Simmons, 1986). The theory of Gerard Ligozat (1994)
for transitions between the endpoints of one-dimensional intervals, provides an elegant representation of these interconnections as a partitioning
of the 2-D state space.

However, even in one dimension, some of the distinct characteristics
of spatial reasoning are obvious. In both space and time the interval
position has continuous transitions from "before" to "meets" to "overlaps." Time, however, does not encourage relative motions of its events,
whereas in space this is critical, and is the basis of all size or distance
comparisons. Time is anisotropic (forward is special), whereas space is
generally isotropic (no global orientation is special).

In Fig. 1.10, the transition between intervals is shown as one interval
moves ahead of the other. In time, these transitions are merely indicators
of "neighboring" relations, such as may be the result of a composition
operation, say. In space, however, these transitions can be real, as in two
objects moving with respect to each other. Furthermore, space may also
permit deformations in its objects, in which case we may have additional
diagonal links as well ("started by" can transition to "equals").

3.1. The Transition From Time to Space

Both space and time are governed by a notion of scale that defines
tangencies; the event of sitting down at a desk *meets* pulling up the chair
in a certain scale just as surely as the desk *meets* the wall in another. In

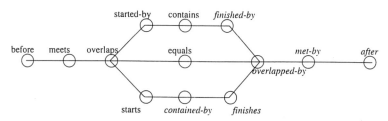

FIG. 1.10. *Interval algebra in time and space:* In both space and time the
interval position has continuous transitions from "before" to "meets" to
"overlaps." Time, however, does not permit relative motions of its events;
whereas in space such placement is critical for size or distance comparisons.
In the transition graph shown, depending on relative size, two intervals
may follow one branch or the other. There are seven primary relations; the
inverse relations are italicized.

both space and time the position relation has continuous transitions of meaning from "before" to "touching" to "overlaps"; although in space, this topology is considerably more complex owing to the greater possibilities opened up by the higher dimensionality.

However, space and time differ fundamentally in several aspects. The greatest difference, of course, is *dimensionality*; what in time is an ordered line becomes in space a multidimensional patchwork. What, for example, is the analog of the notion of perpendicular in one dimension? Or that of a screw in anything less than three dimensions? Dimensionality effects are discernible in the new attribute of *orientation*; but it also shows up in a distinct loss of *transitivity* information, a loosening that leads to much more uncertainty when using the same discretization (−, 0, +, say) in three dimensions than in one. More subtle are the differences in the nature of ordering. Time is *anisotropic*; both thermodynamics and our cognitive experience tell us that going back toward the past is very different from going forward. Space, at least as commonly experienced, suffers no such asymmetry, and is fully isotropic. All orderings, in all directions, are possible: Relations "before" and "after" merge into a bland "disjoint"— thus the seven primary relations are the critical ones; these are the same ones used in topological models (Cohn et al., 1993). Cognitively, space is ordered by our senses, and the *size* of objects leaves a visual memory that is more ingrained than in time. It is easy to compare the height of a man in Bombay with the Empire State Building in New York; it is much more difficult to realize the vastness of the Cenozoic era compared with an individual's life span: The memory of time is essentially nonvisual, concerned more with ordering than with size. Size comparisons actually involve mental operations that superimpose these objects, similar to the "flush" operation where the startpoint of two intervals are aligned to compare their sizes (Mukerjee & Joe, 1990).

Despite these differences, spatial reasoning owes much to one-dimensional interval logic. Much reasoning about space is an attempt to reduce the multidimensionality to several strands of this one-dimensional order. In a well-argued case for this, Ralf Röhrig (1994) developed a theory for qualitative spatial reasoning where he claimed that all spatial reasoning models developed in the AI community so far are just this: projections into some ordered, lower dimensional space. To establish this he developed an ordering for cyclic spaces to handle orientation (see Fig. 1.8), and showed how many of the existing models can be reduced to the projective paradigm. However, the cyclic ordering involves defining a "reference" point on the cycle, and weakens the transitivity power in the ordering. Also, it is possible that some of the models for space that deal with the interaction between angle and position (Jungert, 1994; Mukerjee & Joe, 1990) may not fall within the scope of this generalization. Many, if not most, of the spatial reasoning models, however, do fall into this projective pattern.

3.2. Orthogonal Worlds

A common simplification for higher dimensional objects is that of rectangles (or cuboids) oriented with the axes, an approximation called the *orthogonal world* approximation here. This model is sensitive to the orientation of the global axes in addition to the shape approximation. Slightly more general is the *intrinsic front* approximation, which allows arbitrary orientations (Fig. 1.11).

Cognitively motivated but with possible applications to task-achieving models is the multiscale spatial array model of Glasgow (1993), which describes spatial relations in terms of a variable rectangular grid or lattice. Figure 1.12a shows some of the European countries, maintaining the conceptual spatial relations between them. Figure 1.12b shows the encoding of a desk-and-chair problem in this framework. Due to the global orthogonal frames, an orientable object like a chair is poorly modeled if rotated. On the other hand, for relatively immovable objects, even if the shape is not rectangular, the model can work, as in the map model. This model is simple, and has good cognitive validity. Computationally however, the distinctions achievable in a cellular model are the same as in a projective 2-D model, though the former may be simpler conceptually. Choosing the right resolution to obtain the right level of abstraction seems to be a complex task in such models, because the boundaries disappear or change with scale. The model also claims to support a conceptual hierarchy, but this is not very clear. For example, in the map of Europe (Fig. 1.12a), as the model become grainier, the clusters that are obtained need to remove some of the inner details (*L*-shapes become rectangles, etc.), and it is not clear how this is to be done. Current implementations

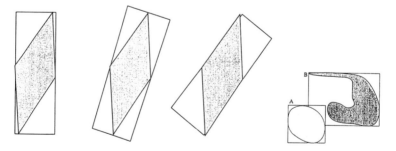

FIG. 1.11. *Rectangular boxing:* Orthogonal world, intrinsic frame, and minimum enclosure models for a diamond shape. The choice of frame, however, is constantly changing during discourse, and modeling these within the rectangular box limitation is difficult. The intrinsic front, where available, appears to provide a reasonable compromise, and is often also minimum enclosure. One weakness of rectangular boxing is that it makes false predictions of intersection.

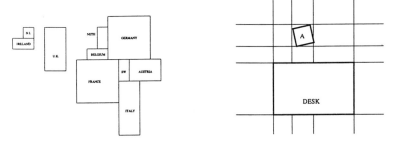

FIG. 1.12. *Spatial array models:* Cognitive models often show an orthogonal orientation, as in this spatial array map of Europe (Glasgow, 1993). For detailed diagrams, the model reduces to a uniform grid. Although such models are quite powerful, some critical parameters, such as the resolution needed, must be obtained through trial and error. Computationally however, the distinctions achievable in a cellular model are the same as in a projective 2-D model (see Fig. 1.6b). On the other hand, the orthogonal world assumption makes it difficult to relate to objects with an intrinsic front, such as the chair.

of this model appear to be arriving at the array sizes through a process of trial and error.

3.3. Objects at Angles

In local frame models, the enclosing rectangle is oriented to the object frame, resulting in a better fit most of the time. However, all rectangular boxing is a poor model for intersection purposes, say. Figure 1.13b shows a combined orientation/position model that uses rectangular boxing with intrinsic fronts. As the chair moves back from A, it remains "in front" longer; at B it is already in the overlap zone, and as it moves back, it

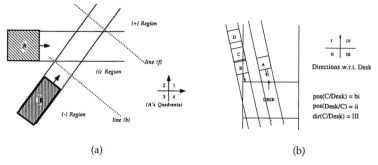

(a)   (b)

FIG. 1.13. *Intrinsic front:* Here the intrinsic front of the object is used to create the rectangle and its projections. Each rectangle defines a front and back (f, b), and three regions (−, i, +). The positions A, B, C, and D, all with the same orientation, correspond to different discretizations in this model; the description for C is shown.

quickly transitions to other qualitative discretizations of the space. Intrinsic front boxing is often identical to the minimum enclosure rectangle, though the long or short sides of the box may be interchanged (e.g., a car vs. a desk). Rajagopalan (1994) used rectangular boxing with intrinsic fronts to reason about a ball rolling down an inclined plane.

A critical factor in local frame models is that of selecting the frame. This may be done depending on the viewer (deictic), with respect to the object itself (intrinsic), or with respect to an external frame (extrinsic). Indeed, recent analysis of speech and dialogue (see chap. 14 of this volume) indicates a number of other modalities such as both-centered, neutral ("between the lamp and the towel"), and a number of others.

Another discretization, based on distance rather than alignment, is the Voronoi diagram, which partitions the space into regions closest to one of a set of points or sites. Such models are widely used in geographical modeling and in robot motion planning. Edwards and Moulin (see chap. 10 of this volume) use such a discretization in building a geographical constraint solver, and also for obtaining mental images by considering weights on the Voronoi regions. A generalized Voronoi diagram for line sites leads to the axis models that have been used in modeling the shape of objects in Agrawal et al. (1995).

3.4. Topology

In one dimension, topology follows directly from the ordering in the real line: Contact is both a positional statement and a topological one. Thus the positional information (e.g., the 13 relations of interval logic) completely determine the topology as well. In two dimensions or higher, shape complexities mean that contact contains almost no positional information, though it may impose some constraints (Fig. 1.14). However,

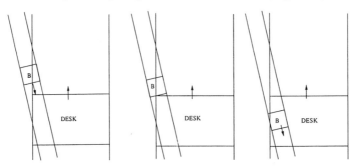

FIG. 1.14. *Qualitative position and topology:* The three configurations shown in the model of Mukerjee and Joe (1990) are all in the qualitative position pos(B/Desk) = ii; pos(Desk/B) = ii; dir(B/Desk) = III. Unlike in one dimension, topological information has to be stored in addition to position and orientation. The same is true for the model of Hernández (1994).

topological reasoning retains the appeal of simplicity (compared with the messiness of position and angle models), and also mathematical elegance.

4. MEASURE-BASED CONTINUUM MODELS

The difficult issue of modeling the complex cognitive connotations of relations such as "left-of" and "near" can be modeled very powerfully using explicit functions, which we call potentials based on the name used in robot motion planning (Latombe, 1991), from which most of these models draw inspiration. Alternatively, continuum measures can also be obtained based on a projective structure, by considering the degree of overlap with different regions, say (Fig. 1.16).

4.1. Potential-Based Measures

Potential-based continuum functions have become popular of late (Gapp, 1994; Olivier et al., 1994; Yamada, 1993). Here one uses fuzzy set distributions, which are referred to as "potential functions," to define the degree or the membership function of a particular configuration in a spatial relation. The essential observation here is that the quality of predicates such as "in-front-ness" is a function of the n-dimensional space, or a field, as opposed to a function of orientation alone, as in projective models. Olivier was interested in interpreting these relations to establish constraints on the placements of objects, which he obtained for relatively nonconflicting placements. Gapp was interested in a more faithful model for the cognitive sense of these terms. Gapp's model can also capture three-dimensional (3-D) notions such as "the aeroplane was over my house." Yamada combined multiple constraints (simply by adding the potential fields) in a database query system.

It is easy to overextend the seemingly endless expressibility of a highly tweakable potential function. For example, Gapp (1994) used two "in front of" functions to obtain a simple model for "between": However, the model fails to capture the fact that the between-ness of an object in the middle decreases as the two reference objects move apart. Furthermore, these functions themselves are determined by choices that are not always easy to justify—they must be mathematically and computationally simple, yet convey the spatial attributes desirable. Finding "good" potential functions remains an art.

A more serious failing in the current batch of potential models is that the located object is modeled as a point, for example, its centroid. This is unrealistic for large located objects—"The library was to the left of me" holds much beyond the centroid of the library (Fig. 1.15). An alternative in this case is to use the configuration space or C-space models (Sacks &

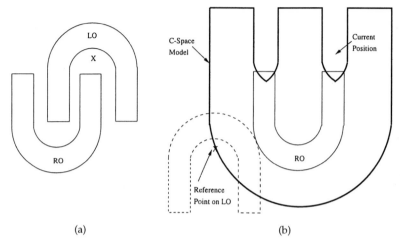

(a)   (b)

FIG. 1.15. *"Two interlocked horseshoes"; Centroid-models fail:* In modeling the
relative positions of two horseshoes, the centroid model of the LO fails to
capture the "interlock" nature of a position, flagging it instead as "in front"
(the X mark in a). This may be better modeled in a potential field function
defined not on the original geometry, but on the configuration space (b),
where the relationship looks very different indeed. The C-space model is
obtained by tracing a reference point (the centroid X) on the LO (dashed
line), as it traces the outer countour of the RO. Positions inside this model
correspond to overlap. The LO has been reduced to a point (the origin) in
the C-space model. The current position as in (a) is now seen to be more
"interlock" in nature.

Joskowicz, 1993; see also section 5.1.1). If the located object LO can only
translate, then the C-space is 2-D as shown, otherwise this C-space may
be considered as the $\theta = 0$ slice in a 3-D C-space (x, y, θ). Although such
a function can indeed handle many different interactions between the
most complex shapes, clearly the problem becomes increasingly complex
and also increases in dimensionality.

4.2. Weighted Sum Over a Discretized Space

Here one uses the overlaps of an object with a discrete set to obtain the
continuum measure. Fuhr et al. (see chap. 6 of this volume) model the
spatial aspects of visual scenes with the spatial prepositions left, right,
front, behind, and above, below. The notion of "degree of fulfillment" is
used for each of these relations in each instance. A number of zones
("acceptance volumes") are defined around each object; for example, in
2-D, there are four projective zones and four diagonal zones; the diagonal
zones are subdivided by a 45° line resulting in a total of 12 regions. In
3-D a similar set of decompositions is arrived at. A relation is defined by

a set of weights assigned to each of these regions. For any given located object, its degree of fulfillment for the given relation is a weighted combination of its overlap with the corresponding regions. This model has the advantage over the potential function models in that the shape of the located object is considered to a certain extent. Just as the function is determined by the programmer in a potential function implementation, the weights here need to be selected very carefully. Furthermore, it is not clear why a particular discretization should be used for acceptance volumes—in Fig. 1.16, the uniform .75 over a whole quadrant may be refinable by subdividing it. Indeed, carrying this subdivision process to its logical conclusion would result in what may be called a weighted potential field—that is, a continuous gradation of the weight.

4.3. Distance Models

Discretizations of near/far and so on have been considered for a long time with the cutoffs set as per the task needs. The number of such categories and their ranges are difficult enough to assign without having to also worry about the aliasing problems that result with points close to the boundary regions. This can be handled better by defining these zones as fuzzy sets, or their close kin, the potential function, as in Fig. 1.9. The problem of reasoning on these models, as in the question of adding

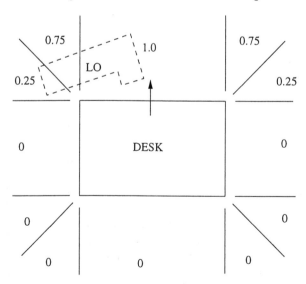

FIG. 1.16. *Weighted sums over an alignment discretization:* The numbers indicate the "in front" relation as a set of weights for the different regions. An object like the dotted outline will have a degree of fulfillment for "in front" determined by the sum of the weighted areas in each region.

distances, was most recently considered in Hernandez et al. (1995) and Mukerjee and Mittal (1995).

A discretization of space based on distance is the Voronoi diagram. An innovative continuum measure defined on Voronoi diagrams was proposed by Edwards and Moulin (see chap. 10 of this volume). Here a Voronoi agent considers the changes that arise when an LO is inserted into the preexisting Voronoi diagram. For example, "near" as a relative measure is captured in the "proximity heuristic"—the agent computes areas lost from older Voronoi regions—proximity is estimated as the ratio of stolen area to original area; if this is large, a good bit of the region that was earlier close to RO is now close to LO, so LO must be near RO. This is a creative approach to defining a continuum measure, and its basis on a distance-based model sets it apart from other approaches. The preprocessing based on Voronoi diagram computation seems to increase the computational overheads compared to a simple potential function, but further work on this model will be needed to indicate its relative strengths compared to the potential function.

5. OTHER ISSUES

Beyond the modalities of discrete versus continuous models, a number of other distinctions animate the discussion of spatial expressions. Foremost among these is perhaps considerations of the shape of the objects. Also important are issues of scale, constraint propagation, and so forth, which are considered in this section.

5.1. Shape

There are surprisingly few models dealing with shape beyond the rectangular enclosure models dealt with in section 2.1(3). One of the earliest approaches to shape modeling—enumeration (Requicha, 1980)—was based on the grid, which is what Glasgow (1993) used in her array model. This also permits scaling by changing the grid tesselation.

Tony Cohn (1995) presented an interesting model for shape using topological concepts together with a convex hull operator and the ability to find adjacent chunks of material from a shape (see Fig. 1.17). This enables distinctions between pointy and smooth projections, collinear/ noncollinear concavities, concavities on same/opposite side, and so forth. This model certainly marks the high-water point for expressiveness based on topological distinctions alone.

In a different approach Agrawal et al. (1995) showed that a class of shapes can be modeled very flexibly by using axis-based definitions as

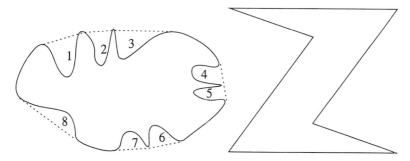

FIG. 1.17. *Shape models: (a) Topological reasoning:* A rich extension of topological models is obtained by adding a convex-hull operation. The differences with the convex hull (recursive) are related topologically to obtain aspects of shape (pointy if touching, as in 2, 3). From Cohn (1995). Copyright 1995 by Taylor & Francis. Reprinted by permission. *(b) Orientation models:* Shape information can also be preserved by storing all signed triangle areas on the vertex set (chirotopes) (Schlieder, 1994).

opposed to contour models (see Fig. 1.7). This type of model can be used in constraint propagation over shapes, where a shape is refined by adding constraints, as in a design task.

5.1.1. Functional Interactions: Configuration Space. We have already seen the effectiveness of configuration space models in considering effects of shape—for example, in the interaction of two horseshoe shapes (Fig. 1.15). But much spatial expression depends on the functional interactions between objects, which can be captured only in the interaction between shapes. We need to focus on the tasks that these shapes do. The efficiency of human models is partly due to the emphasis on a limited, functionally relevant aspect—in "turning the corner of the building" the rest of the building is of little concern. The configuration space model, borrowed from robot motion planning (Latombe, 1991), encodes the interaction between objects in a generalized-coordinates space consisting of the degrees of freedom of both objects. Any one-dimensional line through the C-space encodes the transitions of a set of one-dimensional interval relations. Making this parallel to a rotational degree of freedom results in the relation as seen in Fig. 1.8. Figure 1.18 shows slices in a (x, y, θ) space of a triangle and a rectangle; the coordinate frame is attached at a vertex of the triangle making it "move around" the rectangle, but the same space can be obtained by "moving" the rectangle around the triangle (Mukerjee et al., 1995). Two (x, y) slices are shown; one has an edge of the triangle aligned with a rectangle edge; the other is in general position. Changes in orientation of the triangle move from one edge-alignment configuration to another via a nonalignment configuration. Similarly Fig. 1.15 earlier shows the configuration space for two horseshoe shapes.

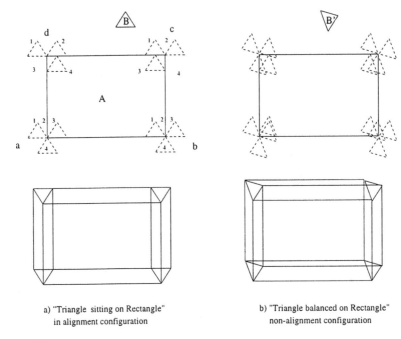

a) "Triangle sitting on Rectangle"   b) "Triangle balanced on Rectangle"
   in alignment configuration   non-alignment configuration

FIG. 1.18. *Configuration space models of object interaction:* Configuration spaces encode the interaction between objects in a generalized-coordinates space consisting of the degrees of freedom of both objects. Two-dimensional motions are encoded in a (x, y, θ) space; two (x, y) slices are shown for a triangle as it "moves around" a rectangle (overlap positions also shown). A line parallel to the theta direction results in the interval algebra model of orientations as in Fig. 1.8. Functional connotations are attached to differences slices; for example, the alignment slice (a) reflects a possible support.

The C-space represents the LO (the triangle in Fig. 1.18) as a point, and identifies all interactions as this "point" moves in this higher dimensional space. Thus, potential models, which typically reduce the LO to a point, would work more smoothly in C-space; furthermore, effects of angular changes, such as that between positions A and B of the chair in Fig. 1.3 cannot be considered except with a C-space model. The trade-off, of course, is in the higher dimensionality and consequent complexity of the model.

Functional aspects can be incorporated as part of the shape interaction, which are nothing but subspaces on the surface of the configuration space model (Sacks & Joskowicz, 1993). For example, the triangle can be supported "on" the rectangle only in alignment configurations like (a), and not in general positions like (b). Similarly, if one of the shapes were a peg and the other a hole, only a part of the C-space surface would correspond to "hole is in the peg." This enables a much broader modeling of spatial expressions in terms of functional context, an aspect that is attracting increasing attention in the processing of spatial expressions (see chap. 16 of this volume). In a sense, some of these issues are also the driving forces behind the transition in the interpretation of spatial prepositions from an "ideal" meaning independent of context (Herskovits, 1986) to a more functionally relevant model.

5.2. Scale and Imprecision

"The book is on the desk." "Pass me the sheet of paper from under the book." The latter statement makes a scale transition from the previous one, where the model initially meant that book and desk touch. Yet at a finer scale, there may be thin objects in the space between them. Modeling statements in different scale spaces, and interrelating them, remains an important open problem, of relevance in spatial domains from Geographical Information Systems (GIS) to visual navigation (Mark & Frank, 1989). Scale and imprecision generally go hand in hand; as the scale (maximum dimensions of model) increases, the tolerance (minimum resolvable distance) increases so as to maintain the volume of data. The effects of imprecision on the boundaries of "neat" discrete set models makes the models less precise, thus tending more toward a continuum model.

An interesting work on modeling imprecision in spatial reasoning is that of Thodoros Topaloglou (1994), who investigates uncertainties or tolerances at the boundaries in a rectangular lattice. This work addresses the issues of scale in the context of orthogonal world projections, where uncertainties are modeled as rectangular tolerance regions called "haze" (Fig. 1.19). This algebra and a similar rectangle algebra can be shown to be complete.

Another model that explicitly deals with multiple scale spaces and transitions between them is that of Mukerjee and Schnorrenberg (1991), which develops transitivity between different tolerance spaces. Given

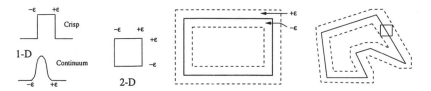

FIG. 1.19. *Tolerances on discrete sets:* The "haze" model extends the 1-D interval imprecision to a 2-D rectangular region. Such a haze can be crisply defined, or it may be a continuum where a membership function defines membership, and not a tolerance. The polygonal shapes on the right have been operated on by a rectangular "haze" operator.

TABLE 1.1
Inference Between Different Scale Spaces

x/y \ y/z	$<_{\epsilon_2}$	$=_{\epsilon_2}$	$>_{\epsilon_2}$	$\leq_{\epsilon_2}$	$\geq_{\epsilon_2}$	$\neq_{\epsilon_2}$	?
$<_{\epsilon_1}$	$<_{\epsilon_1+\epsilon_2}$	$<_{\epsilon_1-\epsilon_2}$	?	$<_{\epsilon_1-\epsilon_2}$	?	?	?
$=_{\epsilon_1}$	$\leq_{\epsilon_1-\epsilon_2}$	$=_{\epsilon_1+\epsilon_2}$	$\geq_{\epsilon_1-\epsilon_2}$	$\leq_{\epsilon_1+\epsilon_2}$	$\geq_{\epsilon_1+\epsilon_2}$	?	?
$>_{\epsilon_1}$	?	$>_{\epsilon_1-\epsilon_2}$	$>_{\epsilon_1+\epsilon_2}$	?	$>_{\epsilon_1-\epsilon_2}$	?	?
$\leq_{\epsilon_1}$	$\leq_{\epsilon_1-\epsilon_2}$	$\leq_{\epsilon_1+\epsilon_2}$	?	$\leq_{\epsilon_1+\epsilon_2}$	?	?	?
$\geq_{\epsilon_1}$	?	$\geq_{\epsilon_1+\epsilon_2}$	$\geq_{\epsilon_1-\epsilon_2}$	?	$\geq_{\epsilon_1+\epsilon_2}$	?	?
$\neq_{\epsilon_1}$	?	$\neq_{\epsilon_1-\epsilon_2}$	?	?	?	?	?
$?_{\epsilon_1}$	?	?	?	?	?	?	?

Note. The entries show the relation between (x, z) given the relations (x, y) at tolerance ϵ_1 and (y,z) at tolerance ϵ_2. Here $\epsilon_1 > \epsilon_2$ without loss of generality.

relations x/y in a tolerance space ϵ_1 and y/z in tolerance space ϵ_2, the inferences for x/z admit normally incompatible relations [e.g., if x/y is < and y/z is =, x/z may be < with the smaller tolerance $(\epsilon_1 - \epsilon_2)$]. This permits the paper to be between the book and the desk, for example. Table 1.1 deals with only one dimension, and can be applied to orthogonal world boxing, as in the haze model, or to scalar results from potential field functions.

5.3. Sensory Clusters and Self-Organization

In modeling and generating spatial expressions, a key issue is the choice of landmarks. Recently, a number of models using sensor data for robot behavior are automating the process of spatial analysis and landmark identification; these also attempt to determine the correct scale for dealing with sensor data. For example, Pierce (1995) managed to identify sensori-motor correlations with objects of interest in the external world, without invoking any a priori knowledge of the learning agent's sensorimotor apparatus or of the structure of its world. Part of this work is identification of the space of action effects, that is, which primitive action vectors cause which motion in each degree of freedom (*dof*). Principal Component Analysis (the determination of the first few eigenvectors from the very large set) is used to diagnose the set of primitive actions. For a 2 *dof* mobile robot with ring sonar, this method discovers the primitive action vectors (−1, 1) for rotation, and (1, 1) for translation. These techniques appear to be very powerful and are likely to find further applications in the near future. Tani (1995) recently claimed this method of sensory clustering as a possible answer to the fundamental symbol-grounding problem of AI—where a system can determine the interpretation of a symbol system intrinsically.

In terms of modeling spatial expressions, such principal component methods may also help in the assigning of frames in a continuously varying model of discourse (see chap. 14 of this volume).

A recent model incorporating velocity and position tolerances tests for collision between two objects (Jungert, 1994). It uses uncertainty rectangles for velocity and orientation—these are similar to the haze rectangles, but have arbitrary orientation, and their positions are qualitatively evaluated by considering only the 12 projection points that result if the vertices of two objects are projected to the axis parallel to one of the boxes. The ordering of these points indicates the likelihood of collision between the two rectangles. Although this is ostensibly a model of uncertainty in motion, the same methodology should prove useful in modeling 2-D angle and position problems as well.

5.4. Constraint Propagation

Given a set of objects and a set of constraints that hold between the spatial positions of these objects, the constraint propagation problem is that of determining either any, all, or an optimum instantiation for positions or *placements* of the objects. A number of models have evolved for discretized spaces. Kramer (1993) modeled the problem in terms of primitive constraints such as those between line, vertex, and arc objects. Bhansali and Kramer (1994) extended this paradigm to solve incremental constraint propagation problems based on geometric knowledge, topology, and a degrees-of-freedom–based data structure. Given a set of geometries (objects) and a set of constraints (e.g., a circle is at distance D1 from line 1), if a new constraint is added (the circle is at distance D2 from line 2), the problem is reduced to that of an agent-based planner, where there is a state description and a number of actions are available to change the state, like rotate, or move. A geometric rule-base matches geometric constraints with the results of actions. Based on this, plan fragments are generated for solving parts of the problem. These are then elaborated, with special measures to remove redundant steps and resolve degeneracies such as extra degrees of freedom, or no solution. An architectural constraint solver (Charman, 1994) considers the problem of placing rectangular regions by using heuristics such as "choose the next-corner-to-be-filled such that the least number of rectangles can be placed there."

In contrast, potential field models represent spatial-prepositional constraints as functions, and interaction between constraints is simply a matter of adding constraints (Olivier et al., 1994; Yamada, 1993). Determining the better solutions now is merely a matter of determining the peaks and valleys in the resulting field, which is however, nonlinear. Local minima may arise though attempts are continuing at refining the potential field definitions to reduce these. These are particularly problematic where there are many constraints over a small region (many things in front of the desk). Representing the geometry through the configuration space instead of making a point-shape assumption, while making the process more robust, exacerbates the nonlinearity of the space, and therefore the local minima problem.

6. APPLICATIONS

Spatial expressions are embedded in the context of tasks, to be performed in certain domains. Whereas neat models attempt generality, the scruffies are content to limit their efforts to certain contexts. Spatial issues that may arise are often treated as peripheral to the task. In the computational or AI context, such tasks have included the blocks world domain, robot path planning and route description, knowledge acquisition from captioned figures or video, and so on. In the four applications selected next, we have tried to provide a flavor of the diversity of approaches in tasks where the spatial aspects appear to be of more significance.

6.1. Route Description

A challenging application that calls for modeling spatial expressions in full generality is that of relating routes or spatial pathways, with verbal directions. This can go from the expression to the map, or from the map to the expression.

Generating route descriptions is an old problem. Kuipers and Byun (1988) considered the problem of finding landmarks in a scene. An early model of bus route descriptions is constructed in Pattabhiraman and Cercone (1989). The salience of landmarks is among the factors considered by Maass (1994), who also used a corpus for incremental route description—for example, instructing a driver while trying to find a place. Gryl and Ligozat (1995) lucidly presented their three-step process for route generation—determine Where to Move (WTM), What to Use as a landmark (WTU), and How to Say It (HTSI). Whereas WTM plans the path, WTU decomposes the path segments for proximity to landmarks (gardens, stations, traffic lights, school, church, etc.). HTSI considers the problem of the points of view of both speaker and addressee (see chap. 15 of this volume) in generating the actual text.

An ambitious attempt in going from the very sparse domain of spatial expressions to the very detailed domain of sketched maps was undertaken by Fraczak (see chap. 11 of this volume). Based on a corpus of path directions, a set of constraints are set up on the spatial entities such as buildings, roads, landmarks, and so forth. Making sweeping default assignments for its objects permits one to generate a first cut at a spatial map—somewhat akin to a back-of-the-envelope doodle while listening to directions on the phone. In the essence, such attempts at reconstructing spatial sketches from sparse linguistic inputs becomes one of marshaling constraints wisely, and keeping options open for reassigning earlier instantiations based on subsequent conflicts.

6.2. Knowledge Acquisition From Captioned Figures

An application of increasing importance in the computational era is that of capturing information from existing media such as documents or video. Linking image objects to entities referred in the caption is a long-standing problem. Abe, Soga, and Tsuji (1981) attempted a line diagram model, and even animated images; Klix and Hoffmann (1979) looked for mental imagery maps; Schirra, Bosch, Sung, and Zimmermann (1987) sought to answer natural language queries based on image sequences; whereas Koons and McCormick (1987) sought to relate neuroanatomy texts to their figures. Recently the CEDAR group at SUNY Buffalo have shown some impressive results with newspaper images. Burhans, Chopra, and Srihari (1996) present results relating to human faces and other objects from newspaper pictures. The spatial aspects are handled using 11 predicates (eight projective, and near, between, and surrounded by). The latter use measures obtained using distances between objects—for example, between occurs if two objects are closer to it than any others. The faces are recognized and labeled from within pictures based on context and caption information. The spatial structure is orthogonal because all documents are usually top-down and left-right (e.g., "Sean Penn, left, is holding a trophy"). Earlier work from the same group (Govindaraju et al., 1989; Srihari & Burhans, 1994) shows the complex knowledge base used in this model with visual hierarchies—lifeform:plant:tree:branch/stem, and so on. To identify the "trophy" held by a person, first the face is identified and then adjacent areas are searched for a "man-made small-to-medium" object.

The reverse process, that of generating related data with graphics images, is a significant tool in database visualization. Mittal, Roth, Moore, Mattis, and Carenini (1995) presented a generator of graphical information that relates items in a database to a graphics image, and generates captions describing the complex display techniques used. An ambitious project for graphical annotation based on a video sequence of a disassembly task is described in Lieberman (1994).

6.2.1. Diagram Understanding. A special class of problems that links images with descriptions and is generating a lot of interest computationally is *diagram understanding*, where parts of the diagram may be labeled and related through functional descriptions (for a review, see Kulpa, 1994). Diagrams are widely used by humans for simplifying spatial reasoning. There does not appear to be a good framework yet for representing diagrams; a popular model is based on the spatial array model of Glasgow (1993). In the work by Narayanan et al. (1994) the task of reasoning about device behavior from diagrams is broken into three phases—effects of nonspatial behavior, effects of spatial behavior on connected components, and determination of component interactions through visualization. Both in this work and also in Olivier et al. (1995), the diagram itself is represented using a spatial occupancy array, which really reduces to a uniform grid or spatial-enumeration approach (Requicha, 1980). Because the author feels that this representation is a poor one for modeling diagrams, this point deserves some clarification.

The problem of diagram modeling is that of identifying the diagram logical element (e.g., lines/boundary segments, or areas/regions). Using a 2-D rectangular lattice has the well-known neighborhood problem— both choices of four and eight neighborhoods result in a topological contradiction where the interior, exterior, and boundary are not consistent. (Other choices, such as six neighborhoods, can detect 45° slopes, but not 135° ones.) Without a stable neighborhood function, it is not clear how such models can be used to detect lines, for example. This would make it very difficult, in our opinion, to segment regions, particularly nonorthogonal lines (e.g., a spring). It may be better to simply represent each line segment as a directed line and relate the endpoints with respect to other lines. Also, some disconnected regions need to have the same label in a diagram (e.g., a diagram region representing a volume is disjointed by a spring).

Once a spatial description can be obtained, diagrammatic analysis is powerful, progressing to visualization by "incrementally modifying the visual representation of the device diagram" (Narayanan et al., 1994). The internal problem representation, together with a large number of domain rules (e.g., behavior of gas, springs, etc.), constitutes a rich qualitative model, on which inferences are made about the behavior of the device. Another application of grid models is in kinematic reasoning (Olivier et al., 1995), where a coarse grid is used to identify parts of the diagram requiring greater emphasis; these parts are then studied with a much finer grid, with an effect similar to that of using a quadtree, say.

Other work on diagrams such as Kook and Novak (1991) or Rajagopalan (1994) (figures used to define physics problems) use simplistic rectangular enclosure and orient the global orthogonal frame by cues (e.g., an arrow along the slope). Though the spatial aspects handled—orthogonal data, no shape information—are simple, the semantic structure generated is rich, partly due to nonspatial attributes (e.g., dot-patterned blocks are magnetic fields). Kook and Novak used a preparsed structure that tags variables with various spatial aspects of the diagram and avoids much of the difficulty in spatial modeling.

7. CONCLUSION

A lot of spatial reasoning ultimately reduces to some sort of constraint propagation or the other. Neat models define discretizations that lead to a discrete programming problem; scruffies reduce these to independent functions that can be added. On the whole, it seems to me that we are at a stage now where the scruffies appear to be winning in the battle for the minds of the spatial reasoning community. In the richness of the model, and in the flexibility of its use, the continuum model has much to offer. However, it is easy to get swept by its power and forget that ultimately determining the parameters for the potential functions (or the weights in a summed-weight model) remains an empirical matter, subject to validation by experiment.

For the spatial reasoning community, important problems that need more attention are in the areas of modeling shape and general multi-dimensional position. Particularly relevant is the modeling of function, without which many spatial expressions may be more difficult to understand. Multidisciplinary solutions such as configuration space techniques (from robotics) may be one of the methods that can throw some light on this. With the richness of results and viewpoints that are emerging from increasing multidisciplinary activity in spatial expressions, more of the deeper problems will become clearer and hopefully more tractable in this critical problem domain.

ACKNOWLEDGMENTS

I would like to acknowledge the input of Raman Rajagopalan, Klaus-Peter Gapp, Daniel Hernández, and many other correspondents and users of my annotated bibliography for their comments.[1] The basic funding for the Robotics Center at IIT Kanpur comes from a number of sources, including the Ministry for Human Resource Development.

REFERENCES

Abe, N., Soga, I., & Tsuji, S. (1981). A plot understanding system on reference to both image and language. In *Proceedings of 7th IJCAI-81* (Vol. 1, pp. 77–84). San Mateo, CA: Morgan Kaufman.

---

[1]This bibliography, with more than 2,500 annotated entries and keyword search, is available at http.cs.albany.edu:/~amit/spatbib.html.

Agrawal, R. B., Mukerjee, A., & Deb, K. (1995). Modeling of inexact 2-D shapes using real-coded genetic algorithms. In P. K. Roy & S. D. Mehta (Eds.), *Proceedings of the Symposium on Genetic Algorithms* (pp. 41–49). Dehradun, India: Bhishen Singh Mahendra Pal Singh.

Allen, J. F. (1983). Maintaining knowledge about temporal intervals. *CACM, 26*(11), 832–843.

Asher, N., & Vieu, L. (1995). Toward a geometry of common sense—A semantics and a complete axiomatization of mereotopology. *IJCAI-95, 1*, 846–852.

Bhansali, S., & Kramer, G. A. (1994). Planning from first principles for geometric constraint satisfaction. *AAAI-94, 1*, 319–325.

Biederman, I. (1990). Higher level vision. In D. N. Osherson et al. (Eds.), *Visual cognition and action* (pp. 41–72). Cambridge, MA: MIT Press.

Björner, A., Las Vergnas, M., Sturmfels, B., White, N., & Ziegler, G. M. (1993). *Oriented matroids*. Cambridge, England: Cambridge University Press.

Buisson, L. (1989). Reasoning on space with object-centered knowledge representations. In A. Buchmann et al. (Eds.), *Proceedings SDD-1*. New York: Springer Verlag.

Burhans, D. T., Chopra, R., & Srihari, R. K. (1996). *Domain specific understanding of spatial expressions*. Unpublished manuscript.

Charman, P. (1994). A constraint-based approach for the generation of floor plans. In F. D. Anger et al. (Eds.), *AAAI Workshop on Spatial and Temporal Reasoning* AAAI, Los Altos, CA (pp. 111–115).

Clementini, E., & Di Felice, P. (1995). A comparison of methods for representing topological relationships. *Information Sciences, 3*, 149–178.

Cohn, A. G. (1995). A hierarchical representation of qualitative shape based on connection and convexity. In *International Conference on Spatial Information Theory COSIT-95*. London: Taylor & Francis. [ftp://agora.leeds.ac.uk/scs/doc/srg/COSIT95.ps.gz]

Cohn, A. G., Randell, D. A., Cui, Z., & Bennett, B. (1993). Qualitative spatial reasoning and representation. In N. P. Carrete & M. P. Singh (Eds.), *Proceedings of the IMACS Workshop on Qualitative Reasoning and Decision Technologies, QUARDET '93* (pp. 513–522). Barcelona: CIMNE. [ftp://agora.leeds.ac.uk/scs/doc/srg/QUARDET93.ps.gz]

Dickinson, S. J., Pentland, A. P., & Rosenfeld, A. (1992). Qualitative 3-D shape reconstruction for 3-D object recognition. In *IEEE Transactions on Pattern Analysis and Machine Intelligence, 14*, 174–195.

Egenhofer, M., & Franzosa, R. (1991). Point-set topological spatial relations. *International Journal of Geographic Information Systems, 5*(2), 161–174.

Forbus, K. D., Nielsen, P., & Faltings, B. (1991). Qualitative spatial reasoning: The CLOCK Project. *Artifical Intelligence, 51*, 417–471.

Forsythe, W. C. (1990). *Fourier analysis for generalized cylinders with polar models of cross-sections*. Unpublished master's thesis, Texas A&M University, College Station.

Freksa, C., & Röhrig, R. (1993). Dimensions of qualitative spatial reasoning. In N. P. Carrete & M. G. Singh (Eds.), *Qualitative reasoning and decision technologies* (pp. 483–492). Barcelona: CIMNE.

Fujihara, H., & Mukerjee, A. (1991). *Qualitative reasoning about document structures* (Tech. Report TR CS 91-010), Texas A&M University.

Gapp, K. P. (1994). Basic meanings of spatial relations: Computation and evaluation of 3D space. *AAAI-94, 2*, 1393–1398.

Glasgow, J. (1993). Imagery: Computational and cognitive perspectives. *Computational Intelligence, 9*(4), 309–333.

Govindraju, V., Lam, S. W., Niyogi, D., Sher, D. B., Srihari, R. K., Srihari, S. N., & Wang, D. (1989). Newspaper image understanding. In S. Ramani et al. (Eds.), *Knowledge based computer systems* (pp. 375–384). New York: Springer-Verlag.

Grigni, M., Papadias, D., & Papadimitriou, C. (1995). Topological inference. In *IJCAI-95* v.1 (pp. 901–907). San Mateo, CA: Morgan Kaufman.

Gryl, A., & Ligozat, G. (1995). Generating route descriptions: A stratified approach. In F. D. Anger (Ed.), *IJCAI-95 Workshop on Spatial and Temporal Reasoning* (pp. 57–64).

Guesgen, H. W. (1989). *Spatial reasoning based on Allen's temporal logic* (Tech. Rep. No. TR 89-049). Berkeley, CA: ICSI.

Hayes, P. J. (1985). The second naive physics manifesto. In J. R. Hobbes & R. C. Moore (Eds.), *Formal theories of the commonsense world* (pp. 1–36). Norwood, NJ: Ablex.

Hernández, D. (1994). *Qualitative representation of spatial knowledge.* New York: Springer-Verlag.

Hernández, D., Clementini, E., & Di Felice, P. (1995). Qualitative distances. In *International Conference on Spatial Information Theory, COSIT-95.* London: Taylor & Francis. [ftp://flop.informatik.tu-muenchen.de/pub/articles-etc/danher.position.ps.gz].

Herskovits, A. (1986). *Language and spatial cognition: An interdisciplinary study of the prepositions in English.* Cambridge, England: Cambridge University Press.

Jungert, E. (1994). Using symbolic projection for spatio-temporal reasoning on tracing objects with uncertain positions. In F. D. Anger et al. (Eds.), *AAAI Workshop on Spatial and Temporal Reasoning* (pp. 101–108). Los Altos, CA: AAAI.

Kaufman, S. (1991). A formal theory of spatial reasoning. In *Proceedings of AAAI Spring Symposium on Logical Foundations of Commonsense Reasoning* (pp. 92–101). Los Altos, CA: AAAI Press.

Kautz, H. A., & Ladkin, P. B. (1991). Integrating metric and qualitative temporal reasoning. *AAAI-91.* Los Altos, CA: Morgan Kaufman.

Kender, J. R., & Leff, A. (1990). Why direction-giving is hard: The complexity of using landmarks in one-dimensional navigation. In *AAAI Workshop on Qualitative Vision* (pp. 195–198). Los Altos, CA: AAAI Press.

Klix, F., & Hoffmann, J. (1979). The method of sentence picture comparison as a possibility for analyzing representation of meaning in human long term memory. In F. Klix (Ed.), *Human and artificial intelligence* (pp. 171–191). New York: North-Holland.

Kook, H. J., & Novak, G. S., Jr. (1991). Representation of models for expert problem solving in physics. *IEEE Transactions on Knowledge and Data Engineering, 3*(1), 48–54.

Koons, D. B., & McCormick, B. H. (1987). A model of visual knowledge representation. In *Proceedings of the First International Conference on Computer Vision* (pp. 365–372).

Kramer, G. A. (1993). *Solving geometric constraints systems: A case study in kinematics.* Cambridge, MA: MIT Press.

Kuipers, B. J. (1977). Modeling spatial knowledge. In *Proceedings of the IJCAI-77* (pp. 292–298). Los Altos, CA: Morgan Kaufman.

Kuipers, B. J., & Byun, Y.-T. (1988). A robust qualitative method for robot spatial learning. *AAAI-88, 1,* 774–779. Los Altos, CA: AAAI Press.

Kulpa, Z. (1994). Diagrammatic representation and reasoning. *Machine Graphics and Vision, 3*(1/2), 77–103.

Kumar, K., & Mukerjee, A. (1987). Temporal event conceptualization. In *Proceedings of the Seventh International Joint Conference on Artificial Intelligence IJCAI-87.*

Latombe, J. C. (1991). *Robot motion planning.* Boston: Kluwer Academic Publishers.

Lieberman, H. (1994). A user interface for knowledge acquisition from video. *AAAI-94, 1,* 527–534. Los Altos, CA: Morgan Kaufman.

Ligozat, G. (1994). Towards a general characterization of conceptual neighborhoods in temporal and spatial reasoning. In F. D. Anger et al. (Eds.), *AAAI Workshop on Temporal and Spatial Reasoning* (pp. 55–59). Los Altos, CA: AAAI Press.

Maass, W. (1994). From vision to multimodal communication: Incremental route descriptions. *Artificial Intelligence Review, 8*(2–3), 159.

Mark, D. M., & Frank, A. U. (1989). Concepts of space and spatial language. In *Auto Carto 8: Ninth International Symposium on Computer-Assisted Cartography* (pp. 538–555).

Mittal, V., Roth, S., Moore, J. D., Mattis, J., & Carenini, G. (1995). Generating explanatory captions for information graphics. *IJCAI-95, 2,* 1276–1283.

Mukerjee, A., Agarwal, M., & Bhatia, P. (1995). A qualitative discretization for 3D contact motions. *IJCAI-95, 1,* 971–979.

Mukerjee, A., & Joe, G. (1990). A qualitative model for space. *AAAI-90, 1,* 721–727. Longer version Texas A&M University CS Dept, TR 92-003, 70 pages.

Mukerjee, A., & Mittal, N. (1995). A qualitative representation of frame-transformation motions in 3-dimensional space. In *Proceedings of the IEEE Conference on Robotics & Automation, Nagoya.* IEEE Press.

Mukerjee, A., & Schnorrenberg, F. (1991). Reasoning across scales in space and time. In *AAAI Symposium on Principles of Hybrid Reasoning.* Los Altos, CA: AAAI Press.

Narayanan, N. H., Suwa, M., & Motoda, H. (1994). How things work: Predicting behaviors from device diagrams. *AAAI-94, 2,* 1161–1167.

Nuallain, S. O., & Smith, A. G. (1995). An investigation into the common semantics of language and vision. *Artificial Intelligence Review, 8*(2–3), 113.

Olivier, P., Maeda, T., & Tsuji, J. (1994). Automatic depiction of spatial descriptions. *AAAI-94, 2,* 1405–1410.

Olivier, P., Ormsby, A., & Nakata, K. (1995). Using occupancy arrays for kinematic reasoning. In F. D. Anger et al. (Eds.), *IJCAI-95 Workshop on Spatial and Temporal Reasoning* (pp. 65–74). Los Altos, CA: AAAI Press.

Pattabhiraman, T., & Cercone, N. (1989). Representing and using protosemantic information in generating bus route descriptions. In S. Ramani et al. (Eds.), *Knowledge based computer systems* (pp. 341–352). New York: Springer-Verlag.

Pierce, D. M. (1995). *Map learning with uninterpreted sensors and effectors.* Unpublished doctoral dissertation, The University of Texas, Austin.

Pratt, I. (1989). Spatial reasoning and route planning using sinusoidal transforms (Tech. Rep. No. TR UMCS-89-8-2). Manchester, England: University of Manchester, Computer Science Department. [http://www.cs.man.ac.uk/csonly/cstechrep/Abstracts/UMCS-89-8-.html]

Rajagopalan, R. (1994). A model for integrated qualitative spatial and dynamic reasoning in physical systems. *AAAI-94, 2,* 1411–1417.

Requicha, A. A. G. (1980). Representation for rigid solids: Theory, methods, and systems. *ACM Computing Surveys, 12*(4).

Röhrig, R. (1994). A framework for comparing theories for qualitative spatial reasoning. *AAAI-94, 1,* 157–159.

Sacks, E., & Joskowicz, L. (1993). Automated modeling and kinematic simulation of mechanisms. *Computer Aided Design, 25*(2), 106–118.

Schirra, J. R. J., Bosch, G., Sung, C. K., & Zimmermann, G. (1987). From image sequences to natural language: A first step towards automatic perception and description of motions. *Applied Artificial Intelligence, 1*(4), 287–306.

Schlieder, C. (1994). Qualitative shape representation. In A. Frank (Ed.), *Spatial conceptual models for geographic objects with undetermined boundaries.* London: Taylor & Francis.

Simmons, R. (1986). Commonsense arithmetic reasoning. Reprinted in D. Weld & J. De Kleer (Eds.), *Qualitative reasoning about physical systems* (pp. 337–343). New York: Morgan Kaufmann.

Srihari, R. K., & Burhans, D. T. (1994). Visual semantics: Extracting visual information from text accompanying pictures. *AAAI-94, 1,* 793–798.

Tani, J. (1995). Self-organization of symbolic processes through interaction with the physical world. *IJCAI-95, 1,* 112–118.

Topaloglou, T. (1994). First-order theories of approximate space. In F. D. Anger et al. (Eds.), *AAAI Workshop on Spatial and Temporal Reasoning* (pp. 47–53). Los Altos, CA: AAAI Press.

Yamada, A. (1993). *Studies on spatial description understanding based on geometric constraints satisfaction.* Unpublished doctoral dissertation, Kyoto University, Kyoto, Japan. [http://cactus.aist-nara.ac.jp/lab/papers/yamada/english/dissertation.ps]

Yamada, A., Yamamoto, T., Ikeda, H., Nishida, T., & Doshita, S. (1992). Reconstructing spatial image from natural language texts. *Proceedings of COLING-92, 4,* 1279–1283.

2

On Seeing Spatial Expressions

Richard J. Howarth
University of Sussex

Since spatial expressions are used to convey information in conversations and other forms of communication, it also seems likely that they are used as part of the mechanism for spatial perception. This chapter describes an approach that implements part of this mechanism to combine two separate exocentric and egocentric representations within a single framework for scene description. The objective of this framework is to support the interpretation of spatial relationships between objects while using the scene as its own best memory.

1. INTRODUCTION

In this chapter we consider some issues associated with how a computer program "understands" visually perceived instances of "spatial expressions" in a scene. This does not directly concern communication but it is a connected issue, dealing with internal knowledge that is associated with those expressions people use to relate their spatial perceptual experiences to one another, when conveying the size, shape, orientation, and position of objects. After all, to communicate details about a spatial perceptual experience we need to have some understanding of what was perceived and be able to verbally describe our perceptions.

It is difficult to see which comes first: knowledge of how to use spatial expressions in conversation, or their use in observation. In this chapter we do not consider how this knowledge is learned (where the perceiver sees how other people use spatial expressions); instead we try and describe what this knowledge is and how it is used by the active observer trying to understand the scene.

The problem we consider is how to express the spatial relationship of objects in the scene that the official observer (i.e., a camera) perceives without building and maintaining a complete three-dimensional (3D) dynamic description of the environment. To address this we pull together background material on qualitative representation, frames of reference, and attention. From this we identify the need for an approach that combines two separate exocentric and egocentric representations within a single framework.

The work described here grew out of research on high-level vision concerned with the surveillance of wide-area scenes (see Howarth & Buxton, 1992, 1993; also see chap. 5, this volume). Here we use some of the techniques developed for the dynamic surveillance applications and illustrate how they operate using less complex static examples. By removing the dynamic component we are attempting to make the exposition of spatial expressions clearer, although it does mean that nothing changes in the scene, that is, no events and no observable behavior, which are key components of surveillance. In the surveillance application, where traffic traversed a roundabout, part of the problem concerned describing the various spatial relationships that hold (and change) between vehicles.

2. BACKGROUND

There are numerous ways of representing knowledge about space, the selection of which depends on the application domain, and the intentions and objectives of the project. For example, in computer vision it is tempting to think that once the location of the various scene objects are obtained in three-space (to some precision with possible errors) then the job of spatial representation is complete. However, this is not necessarily the case. For example, in surveillance we still have the problem of understanding what the recognized objects are doing. One of the first things to consider concerns the general form of representation to be used and whether a qualitative or quantitative representation is the most suitable. We follow these considerations with discussions of frames of reference and attention.

2.1. Quantitative Versus Qualitative

Reasoning with quantitative data is not always easy, see for example Paul's (1981) book on robot manipulators. Mukerjee and Joe (1990) also discuss this subject, identifying how a qualitative representation can provide a trade-off between precision/detail and flexibility/abstraction and how a qualitative approach can provide an analogical mapping that inherently reflects the structure of the represented domain. Additionally a qualitative representation is appealing because it maps well to natural language, and to how humans judge and express measurement, and is able to provide a general, more abstract, answer without bothering with specific details. However qualitative models lack rigor and are context dependent and because of this potential ambiguity can be difficult to understand.

A qualitative representation can still be based on precise measurement data; it is just that these data are not used in a quantitative way, instead we use the qualitative representation to provide a useful abstraction that makes clear important data. Kalagnanam, Simon, and Iwasaki (1991) described how qualitative reasoning treats quantities that are scaled only ordinally. Hence they are defined only up to arbitrary, strictly monotonic,[1] transformations providing topological boundaries between these ordinal terms. "More" and "less" describe qualitative relations between pairs of values of corresponding ordinal variables, for "amount" in this case.

There are a number of qualitative spatial representations, such as those described by Freska (1992), Hernández (1991), Mukerjee and Joe (1990), each of which describes an interesting compositional model.

Mukerjee and Joe (1990) extended a one-dimensional (1D) representation of temporal intervals, which has an implicit single direction, to provide a two-dimensional (2D) representation of space carrying over this assumption to provide a representation that integrates topological concepts and motion. Table 2.1 illustrates the additional decomposition this has provided.

Part of Hernández's (1991) paper describes a qualitative spatial representation that can be used to model orientation (with eight ordinal values) together with what he called "projection" (which models a subset of the spatial relationships described by Egenhofer & Al-Taha, 1992, and Egenhofer & Sharma, 1993) having the ordinal values listed in Table 2.1. Hernández detailed transitivity tables for reasoning about their combinations by using constraint satisfaction. The representation mixes spatial and abstract data together to provide a slightly confusing graphical notation.

Freska (1992) discussed some of the existing approaches to qualitative spatial reasoning and introduced the notion of qualitative orientation using an iconic representation that can be composed. For example, if we know the orientation of c given vector **ab** and we know the orientation of d given **bc**, Freska showed how to determine d with respect to **ab**.

---

[1]If (S, α) and (T, β) are partially ordered sets where S and T are sets, and α and β are binary relations, then a strictly monotonic function f has the property that if $x\alpha y$ and $x \neq y$ then $f(x)\beta f(y)$ and $f(x) \neq f(y)$. (See, for example, Barr & Wells, 1990, for more details.)

TABLE 2.1
A Comparison of the Ordinal Values Used by Egenhofer
(Egenhofer & Al-Taha, 1992; Egenhofer & Sharma, 1993)
With Those From Hernández (1991) and Mukerjee and Joe (1990)

Egenhofer	Hernández	Mukerjee and Joe
disjoint	disjointness	ahead
		posterior
meet	tangency	front
		back
equal	(mutual/perfect) overlap	interior
overlap	overlap	
covers		
coveredby		
contains	contains	
inside	included	

An important element of these representations is the frame of reference of the objects to which the representation is applied, and that we discuss next.

2.2. Models for a Frame of Reference

Frames of reference are used in both quantitative (see Paul, 1981) and qualitative representation (see section 2.1). There are basically three frames of reference that we can use in our spatial representation: the environment-centered frame, the object-centered frame, and the perceiver-centered frame (see Carlson-Radvansky & Irwin, 1993). Perhaps the most obvious one is the environment-centered frame (at least from the Western viewpoint; see Hutchins, 1983), which uses a global coordinate system around some selected origin, enabling us to model the world in a "maplike" way,[2] and model the position of an object as a centroid or by its spatial extent. This sort of environmental representation is useful for modeling large-scale space (Kuipers & Levitt, 1988) and communicating spatial information in terms of diagrams, say as part of a discussion about rearranging furniture in an office. Although this is undoubtedly a good representation for some things, we can simplify this further by using an orthogonal quadrilateral representation (see Wood, 1985) with cuboid objects that align their planes with the coordinate axis (e.g., Hernández, 1991). However, although this may work to some extent in our office reorganization example (introduced earlier), where we are dealing with cuboid shapes like desks, it is a too well

---

[2]Or more correctly, establishing a homeomorphism between a portion of the surface of a sphere and a portion of the Euclidean plane.

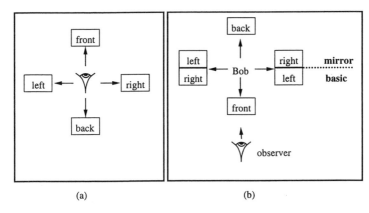

FIG. 2.1. Frames of reference: (a) an observer in "canonical position," (b) the "canonical encounter." Adapted from Herskovits (1986). Copyright 1986 by Cambridge University Press. Adapted by permission.

regimented representation that is not capable of describing the shape or orientation of all objects in the office (i.e., round or curved objects and those not aligned with the coordinate axes). As Mukerjee and Joe (1990) pointed out, if we try and make these nonconformist objects comply with this representation by giving them imaginary rectangular enclosures (Cameron, 1989, called these "surrounding bounding boxes") then there is also the problem of nonaligned objects having rectangular enclosures that overlap when the actual objects do not. If we are not using an orthonormal constraint, describing the orientation and position of one object in terms of another can be complicated in an environment-centered representation.

To reduce this complexity we could use the object's frame of reference, which involves (as observed by Ballard, 1991) fixating on the point in the scene where the object is and pinning an appropriate exocentric coordinate frame to it. Herskovits (1986) described how a human being learns to construct a frame of reference starting from two basic experiences: (a) the experience of looking straight ahead with his or her body standing upright on horizontal ground (we call this the "canonical position"), and (b) the experience of encountering another human being face-to-face (the "canonical encounter"). In Fig. 2.1(a) we see the horizontal plane with the observer (denoted by an eye shape) looking forward. The horizontal axes are described by terms like left/right, front/back/side, before/behind.[3] In Fig. 2.1(b) the perceiver in effect "combines" the point of view of the person encountered (Bob in the figure) with his or her own. The front and back axes are Bob's and point in directions opposite to those of the

---

[3]In this chapter we do not consider the vertical axis and do not use terms like top/bottom, above/below, over/under, even though this information might be available to the official observer.

onlooker. However, the right and left axes can have either the same direction as the observer's right and left, called "mirror order," or the opposite of Bob's right and left, called "basic order."[4] As shown in Fig. 2.3, this conflict between mirror and basic order can raise problems when interpreting the behavior of observed actors/objects in the environment. In Howarth and Buxton (1992, 1993), we use basic order, so in the case of Fig. 2.3, we would say that the van is making a right-hand turn. We have now described the three frames of reference introduced at the beginning of this section: the maplike environment-centered frame called "extrinsic," the object-centered frame called "intrinsic," and the perceiver-centered frame called "deictic"[5] and that is used by the official observer.

The official observer describes the scene objects via their (i.e., the scene objects') own frame of reference (i.e., the frame of reference that the observer, looking at them, assigns to the objects). To do this assignment, an object generally needs an intrinsic front (determined by knowledge about the object and/or its movement). However, it is not aways easy to identify the intrinsic front of an object because some objects such as natural ones like rocks, clouds, and plants might not have a distinguishable front, in which case we assign the intrinsic front to either that part of the object facing the observer or the leading face defined by its motion.

Now that we have the frame of reference we can use it to interpret the space local to the object perhaps using qualitative models like those illustrated in Fig. 2.2. In the figure, part (b) decomposes space into four quadrants to describe front, left, back, right; part (c) makes more explicit use of the objects extent using four half-planes to partition space into eight sections (for more details see Howarth & Buxton, 1993); and part (d) is based on Hernández's (1991) orientation model. Mukerjee and Joe (1990) also described models similar to parts (b) and (c) and gave transitive relationships for them combined with their set of ordinal values given in Table 2.1. To adopt such an object-centered approach means that we are no longer using just an environment-centered frame.

2.3. Attention

The official observer does not need to look at all the objects in the scene; instead it only needs to attend to those relevant to its current perceptual task. The problem of selecting important figures is reviewed by LaBerge

---

[4]Hernández (1991) called this "tandem order" (p. 72).

[5]Deixis is used in several disciplines such as linguistics (Bühler, 1982; Levinson, 1983), the social sciences (Garfinkel, 1967; Heritage, 1984), and spatial representation (Herskovits, 1986; Retz-Schmidt, 1988). Deixis is the use or referent of a deictic word (e.g., I, now, this, that, here) and is an aspect of a communication whose interpretation depends on knowledge of the context in which it occurs.

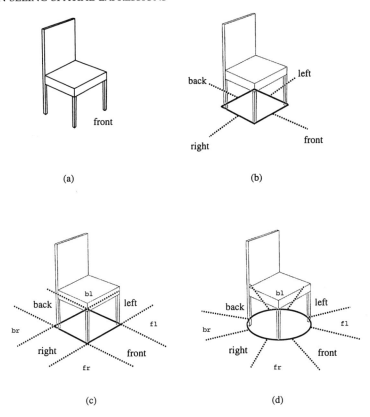

(a)   (b)

(c)   (d)

FIG. 2.2. Example spatial partitionings.

(1990), and described by both Rock (1983) and Treisman (1985), who proposed attributing visual processing to two stages called "preattentive" (or peripheral) and "attentive" (or foveal). Pylyshyn and Storm (1988) also used this separation in their FINST theory by separating multitarget visual tracking into two stages, one a parallel preattentive indexing stage and the other a serial checking stage invoked in selecting a response. At the first stage simple features are preattentively registered that describe global features but not fine detail. Treisman provided examples of texture segregation (a prerequisite for figure–ground separation). Murray, McLauchlan, Reid, and Sharkey (1993) discussed other cuelike processes such as one to detect "looming motion."

At the second stage objects are identified using the candidates set up by the preattentive stage. Mahoney and Ullman (1988) described how the results from demonstrations of the preattentive stage support their use for directly indexing local features as long as the figure of interest is distinguished from irrelevant figures by a single one of these features. In this way preattentive processing can propose figures for use by attentive processing. For example, once the first stage has identified a contiguous blob of space, we can mark this location and bring specialized processing to bear on the target. This may involve things like working out what it is, by first attending to the whole object then adjusting downward to align with parts of the object.

In the first stage some of the object interrelationships that we wish to describe are similar to the early discoveries made in Gestalt psychology (see, e.g., Gordon, 1989) concerning grouping properties, such as proximity, similarity, good configuration, common fate (spatial and/or temporal continuity), closure, and symmetry.

3. SEEING SPATIAL EXPRESSIONS

We have now introduced qualitative representation, frames of reference, and attention. In this section, we draw on these three to develop an approach that enables a perceiver to describe spatial expressions using object-centered frames and resolves the conflict between mirror and basic order.

3.1. Local and Global Representation

From the three frames of reference described in section 2.2 we can form two useful definitions:

> *Definition 1.* The *local form* is representation and reasoning that uses the intrinsic frame of reference of a perceived object (exocentric with respect to the observer).
>
> *Definition 2.* The *global form* is representation and reasoning that uses the perceiver's frame of reference, which operates over the whole field of view (egocentric with respect to the observer).

The global form is not a public world because it, like the local form, only exists to the perceiver. We are not representing a shared world in terms of each participant.

In section 2.2 we encountered an overlap between mirror order and basic order. In Fig. 2.1(b) the front and back axes are Bob's and point in directions opposite to those of the onlooker. However, the right and left axes have the same direction as the observer's right and left, the opposite of Bob's right and left. This symmetry makes left/right distinctions hard to learn and even adults can confuse the two. Herskovits (1986) said that the difficulty would probably be overwhelming if, besides drawing the distinction correctly on themselves, speakers and hearers had to reverse right and left. However, basic order is more useful because it does not cause problems with perceived objects that change between facing and not facing the observer. Bühler (1982) also mentioned this basic order model when he described a gymnastics teacher facing a dressed line of gymnasts and giving commands where the orders *left* and *right* are conventionally given and understood according to the gymnasts' orientation. Bühler noted the astonishing ease of translating all field values of the visual system and the verbal deictic system from someone in another plane of orientation. Bühler's description appears to be at odds with Herskovits'. However, we can resolve this by referring to Fig. 2.3 and identifying that Herskovits was using the perceiver's frame of reference and that Bühler was using the intrinsic frame of reference. The difference is between "global" and "local" reasoning; that is, mirror order is using the global form (definition 2) and basic order is using the local form (definition 1). Using mirror order the van in Fig. 2.3 turns left; if we use basic order then the van turns right, making a right-hand turn. This access to the frame of reference of a perceived object allows us to analyze all the changes involved or transpositions through space of such an object in terms of our own field values and experience. This provides a more *natural* feel because it is the system that we use in our everyday interactions with the world.

The local and global representations are different in terms of granularity and use. They complement each other, although additional reason-

FIG. 2.3. Does the van turn left or right if it follows the arrow?

ing is needed to combine these two approaches. The global form provides a conceptual framework within which to locate (index) locally derived information and because of this it only requires a coarse representation of the scene that can support its dual role of retaining indexical data and holding preattentive details about the scene.[6] In contrast, the local form represents attentive details about the currently attended object and the space local to it (such as the data accessible via the qualitative relative positions described in Fig. 2.2). Neither of these representations replaces or fully duplicates the real world that is perceived by the observer.

This has introduced the local and global representation used by the official observer showing how they are similar to the attentive and pre-attentive stages described in section 2.3 with the local form being a superset of attentive processing. To illustrate this we expand on the deictic representation introduced earlier.

3.2. Deictic Representation

We begin by introducing the terms *entities* and *aspects*, which are used by Agre and Chapman (1987) and Chapman (1991) to describe two important elements of indexical reference, that is, "where is it" and "what is it," such that we have to know where an object is in the scene before we can find out what the object is:

> *Definition 3.* An *entity* is something that is in a particular relationship to the agent.
>
> *Definition 4.* An *aspect* describes a property of an entity in terms of the agent's purpose.

In this chapter the agent is typically the official observer.

Aspects provide information about the current "states" of an object. For example, *the-mug-I-am-drinking-from* is the name of an entity and *the-mug-I-am-drinking-from-is-hot* is the name of an aspect of it; although, outside the remit of this chapter, it is worth pointing out that each time we obtain[7] an aspect (say by picking up *the-mug-I-am-drinking-from*) its value may be different (e.g., as *the-mug-I-am-drinking-from* cools). Also the entity is not

---

[6]For example, initially this would be a collection of precategorical segmentations (blobs) and remain that way until the official observer attends to each one, exerting further visual processing, and once attended to details about the object are remembered. Typically, we only notice these processes when we have difficulty understanding an image (such as looking at a painting where we sometimes miss details), particularly one that is ambiguous.

[7]By the word *obtain* we are finessing the underlying machinery that uses an input sensor and converts the result into a value that represents some property value in the current state of the world.

always the same (e.g., during a typical day, on different occasions, I drink from a number of different mugs). However, there is some local temporal continuity which, most of the time, solves this problem.

When reasoning about an object in the scene, it is useful to use the object's local form (see definition 1) to obtain aspects about the other objects in relation and relative to the attended object. This includes such things as the position of another object in relationship to the van in Fig. 2.3, or the relationship of the phone to the mug in Fig. 2.4, for example, *the-mug-I-am-drinking-from-is-next-to-the-phone*, *the-mug-I-am-drinking-from-is-infront-of-the-phone*, and *the-mug-I-am-drinking-from-is-beside-the-phone*. We can also describe the spatial relationship between the mug and the mat: *the-mug-I-am-drinking-from-is-sitting-on-a-beer-mat*. All these deictic spatial expressions obtain information about the scene by using an aspect of an entity (i.e., indexing an object and obtaining information from its local form). A problem here is integrating the information from two spatially related but separate objects such as the phone and the mug. Representing and reasoning about this spatial information, initially at least, does not appear complex. However, developing an implementation brings to light unexpected problems concerning the integration of disparate perceptions due to providing each local scene object its own frame of reference (orthogonal coordinate system). One way of doing this is to make additional use of the global form. In the first place it acts as a mechanism for maintaining spatial indexes, and in the second place it coordinates the composition of local data.

Interestingly, part of the background to this work on deictic representation comes from the work of Garfinkel (1967) and Heritage (1984) on communication, who used the "reflexive accountability" employed by each participant to interpret the perceptual data, to both fill in the context

FIG. 2.4. The phone.

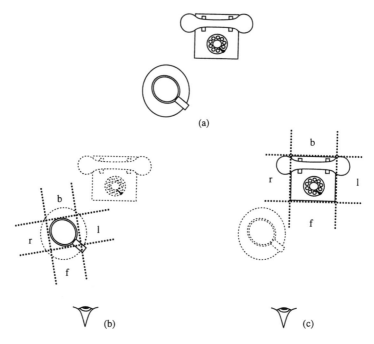

(a)

(b)   (c)

FIG. 2.5. Each object has a local model with respect to the observer.

and identify abnormal behavior (breaches of the norm). Although we are just considering the official observer in this chapter, we assume that it is applying the same mechanisms to interpret the behavior of agents in the scene.

3.3. Implementation

When we consider how to implement this local and global model we come across the problem of selecting suitable coordinate systems. To simplify things we just consider a single fixed camera, a "cyclopean model," with one image plane which, when combined with suitable knowledge, can be modeled as a 3D world and transformed to a ground-plane representation to facilitate ease of expressing spatial relationships in the horizontal axis[8] (as shown in Figs. 2.5 and 2.6), although the image plane would work just as well.

In Fig. 2.5, we are modeling the complex process of visual attention using mutual proximity as an attentional cue for pairwise relationships

---

[8]This approach originated from work on road-traffic surveillance (see Howarth & Buxton, 1992, 1993) that performed spatial reasoning in the ground-plane.

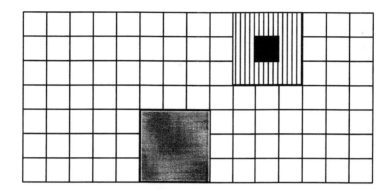

FIG. 2.6. Each differently shaded area represents an object which is part of the observer's global model.

(see Howarth & Buxton, 1993, for more details). The eye shape in the figure represents the observer, which in an implementation would be a camera connected to a computer-vision system with (a) a *preattentive stage* that identifies, at a coarse level, where the objects/blobs are, and (b) an *attentive stage* that recognizes what the foveated object is and its pose in the environment (say, by using a model-matcher). After running the preattentive stage, we then attend to each blob/object that the preattentive cue has identified as likely to be interesting or important in some way. In this implementation, to attend to the selected object, we perform model-matching, so we can identify the object's front, if present, and attach an appropriate local frame of reference. In Fig. 2.5b we have done this for the mug. Using the mug's frame of reference, we can say there is some object behind and to the left of it (generating the aspect *the-other-is-behind-and-to-the-left*). Then we attend, by performing model-matching again, to this other blob that turns out, as shown in Fig. 2.5c, to be the phone. This visual operation does not remember the previous model-matcher results, it just performs a recognition process on the current attended object. We can attach another local frame of reference to *the-phone* and index the blobs proximate to *the-phone* in terms of this. So when attending to *the-phone* we get the aspect *the-other-is-infront-and-to-the-right*.

The problem is how to combine these two deictic descriptions (we will call this deictic data fusion the *viewpoint integration* problem) and to do this we use the global model (in Fig. 2.6, this is depicted as a grid). This grid is just to simplify implementation details—any coarse representation would do. This is because in the global form, object detail is not so important; all we need to represent is some coarse positional information. As shown in Fig. 2.6, we can use the global model to represent the locations of the unrecognized blobs.

This coarse grid also provides an environment-centered coordinate system for the indexical reference of scene objects. Although simple, this model is similar to the preattentive "winner takes all" technique of Koch and Ullman (1985; also see Chapman, 1991), who presented cognitive grounds for this approach. In Fig. 2.7, we illustrate how this indexical framework is used to (a) link global positional information of unrecognized blob locations to results from attentional processing (such as model-matching just introduced); (b) select which objects to attend; and (c) enable indexing of aspects about each object in the real world, such as its type, actual size, and so forth, or the relative positional aspects of nearby objects.

To solve the viewpoint integration problem involves making sense of these relative positional aspects. To do this the official observer integrates results from each object's local frame of reference to provide a single "global" description. In Howarth and Buxton (1993) the implementation uses Bayesian networks to model the uncertainty present in dynamic visual images, although other approaches that can represent uncertainty would probably operate just as well (e.g., fuzzy logic). To illustrate how this deictic information is composed let us look at Fig. 2.6, which shows two situations where we attend to *the-mug* and *the-phone*. As just described, if we attend to *the-mug* then *the-other-is-behind-and-to-the-left*, and if we attend to *the-phone* then *the-other-is-infront-and-to-the-right*. We can compose these with respect to Table 2.2, where to generalize the model we use *the-other* as the name of the indirectly indexed object and *the-refobj* as the name of the attended object. In Table 2.2 we have a set of statements

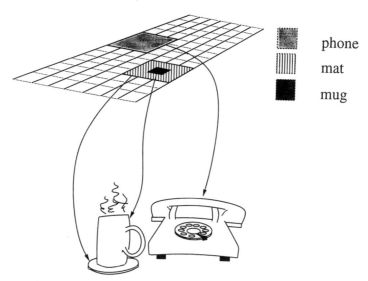

phone

mat

mug

FIG. 2.7. Indexical framework that is used to identify the global-form "blobs" and their respective spatial relationships.

TABLE 2.2

A Subset of the Composition Rules

VIEWPOINTS FROM EACH *the-refobj*			COMPOSITE	
			locative	*motion*
other-is-behind	other-is-behind	→	back-to-back	back-to-back
other-is-behind	other-is-infront	→	one-behind-the-other	following
other-is-right	other-is-right	→	rightside-adjacent	pass
other-is-left	other-is-right	→	facing-the-sameway	side-by-side
other-is-infront	other-is-infront	→	head-on	head-on
other-is-behind-and-to-the-left	other-is-infront-and-to-the-right	→	facing-the-same-way, staggered/diagonal	same-heading, staggered

of the form "*a b → c d*" where *a* and *b* each describe a viewpoint from the frame of reference of two different *the-refobjs* such that we want to integrate these mutual descriptions of one another into one overall description. There are two interpretations we can make dependent on whether the objects/actors involved are stationary or moving, we call these "locative" and "motion" respectively, and in Table 2.2, these correspond to columns *c* and *d*. In Fig. 2.8 the table to the right illustrates the number of combinations that integrating the viewpoints from two *the-refobjs* can take. These spatial arrangements of two objects can be interpreted as describing two static figures, or as being "snapshots" of two moving objects. In the figure the shaded column describes the various positions of *the-other* (which at this stage is just perceived as a contiguous blob, so *the-refobj* knows that something is there but not what it is) in relation to the shaded *the-refobj* with the arrow denoting *the-refobj*'s front. Similarly for the unshaded one. The smaller part of Fig. 2.8 on the lower left describes the positions that *the-other* can take in relation to *the-refobj* that we are attending.

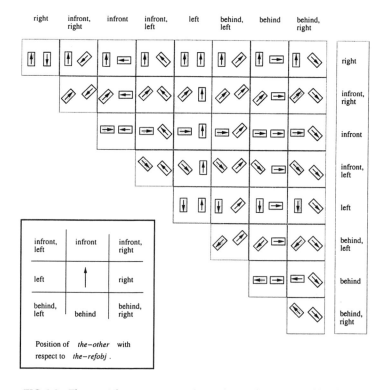

FIG. 2.8. The spatial arrangements of two objects shown as a table of icons on the right with *the-refobj*'s frame of reference shown on the left.

We can use the results from Table 2.2 to describe the spatial relationship that holds between *the-mug* and *the-phone*. Under the locative interpretation the results are `facing-the-same-way` and `staggered/diagonal` reflecting their diagonal spatial relationship. Alternatively if we were to apply these rules to describe similarly positioned objects in the road-traffic domain, where we need to take account of dynamic scene objects, then under the motion interpretation the results would become `same-heading` and `staggered`. We can also use the two deictic descriptions of *the-other* to look-up the corresponding icon in Fig. 2.8 that represents the same result. Issues, related to those depicted in Fig. 2.8 and Table 2.2, concerning (a) motion are described by Talmy (1983, p. 255), and (b) spatial extent are discussed by Herskovits (1986, p. 187) where she considers spatially extended located objects.

A problem with this compositional model is that in fusing the relative positional results we lose information concerning which individual object contributed what result. This can be resolved by reattending to the objects in the scene and, if necessary, performing more complex spatial reasoning, such as by using the visual routines described by Chapman (1991), which manipulate attentional markers, rays, and activation planes.

Hernández (1991) described a reference frame model similar to the one used here (see Fig. 2.2 for a visual comparison) and detailed how spatial relations can be composed, noting how composition leads to serious degradation in detail although this degradation can be reduced by reference to more scene objects. To obtain an internally consistent representation of the world, Hernández proposed constraint propagation, something not used here because to find out where some object is, it is best to just look at the object (and assume that the world is consistent [most of the time]).

4. CONCLUSION

In this chapter, we have described a framework that integrates egocentric and exocentric representations. This provides spatial reasoning that can make use of each scene object's local frame of reference to obtain details about the object and its surroundings, and incorporate these results into a more global model so that the results from attending to objects provide not just details about the individual objects but also about their mutual spatial relationships. We have shown that, in this framework, there can be more than one spatial representation and that these representations can complement each other.

The approaches described here are for representing perceived information about the world, and providing a representation for indexing every piece of data in the world without having to make a complete internal copy of it. These representations are more useful in a dynamic context, with moving objects, than the static one described here, where the preattentive information can include data about the velocity of objects and Fig. 2.8 can be interpreted to describe and identify how episodes like *following* and *overtaking* evolve (for more details, see Howarth & Buxton, 1993). Naming each ordinal value in the set of spatial arrangements in Fig. 2.8 is hard, perhaps illustrating a limit on how we express spatial information via natural language; the iconic model is so much easier. This problem was eased in Howarth and Buxton (1993) by assuming that the two agents had a similar heading because they were near each other and on the same side of the road.

If we had an active camera platform (such as the one described by Murray et al., 1993), then the global form could act as a short-term memory of where objects are so that the official observer can look away from an object and then look back, having some idea of where the object should be. Also the 2D global form could be extended to a 3D model so that it can also hold information about depth (or even some form of 4D model holding motion vector information).

ACKNOWLEDGMENTS

The work reported on in this chapter was funded at various stages by SERC under a CASE award with GEC Marconi Research Center, and by the EPSRC project "Behavioural Analysis for Visual Surveillance Using Bayesian Networks" (Grant GR/K08772).

I thank Annette Herskovits for giving me permission to base Fig. 2.1 on Figs. 10.2 and 10.3 from pages 158–159 of her book (Herskovits, 1986). Also, my illustration in Fig. 2.1 is reprinted from Fig. 4 of Howarth (1995) with kind permission from Kluwer Academic Publishers.

REFERENCES

Agre, P. E., & Chapman, D. (1987). Pengi: An implementation of a theory of activity. In *Proceedings of the Sixth AAAI Conference* (pp. 268–272). San Mateo, CA: Morgan Kaufman.
Ballard, D. H. (1991). Animate vision. *Artificial Intelligence, 48*, 57–86.
Barr, M., & Wells, C. (1990). *Category theory for computing science.* Englewood Cliffs, NJ: Prentice-Hall.
Bühler, K. (1982). The deictic field of language and deictic words. In R. J. Jarvella & W. Klein (Eds.), *Speech, place, and action: Studies in deixis and related topics* (pp. 9–30). New York: Wiley. [An abridged translation of *Sprachtheorie*, Jena: Fischer, 1934].
Cameron, S. (1989). Efficient intersection tests for objects defined constructively. *The International Journal of Robotics Research, 8*(1), 3–25.

Carlson-Radvansky, L. A., & Irwin, D. E. (1993). Frames of reference in vision and language: Where is above? *Cognition, 46*(3), 223–244.

Chapman, D. (1991). *Vision, instruction and action.* Cambridge, MA: MIT Press.

Egenhofer, M. J., & Al-Taha, K. K. (1992). Reasoning about gradual changes of topological relationships. In A. Frank, I. Campari, & U. Formentini (Eds.), *Theories and methods of spatio-temporal reasoning in geographic space* [Lecture Notes in Computer Science 639] (pp. 196–219). New York: Springer-Verlag.

Egenhofer, M. J., & Sharma, J. (1993). Topological relations between regions in $\Re^2$ and $\mathbb{Z}^2$. In D. Abel & B. Ooi (Eds.), *Advances in spatial databases* [Lecture Notes in Computer Science 692] (pp. 316–336). New York: Springer-Verlag.

Freska, C. (1992). Using orientation information for qualitative reasoning. In A. Frank, I. Campari, & U. Formentini (Eds.), *Theories and methods of spatio-temporal reasoning in geographic space* [Lecture Notes in Computer Science 639] (pp. 162–178). New York: Springer-Verlag.

Garfinkel, H. (1967). *Studies in ethnomethodology.* Englewood Cliffs, NJ: Prentice-Hall.

Gordon, I. E. (1989). *Theories of visual perception.* New York: Wiley.

Heritage, J. (1984). *Garfinkel and ethnomethodology.* Cambridge, England: Polity Press.

Hernández, D. (1991). Relative representation of spatial knowledge: The 2-D case. In D. M. Mark & A. U. Frank (Eds.), *Cognitive and linguistic aspects of geographic space* (pp. 373–385). Boston: Kluwer Academic.

Herskovits, A. (1986). *Language and spatial cognition: An interdisciplinary study of the prepositions in English.* Cambridge, England: Cambridge University Press.

Howarth, R. J. (1995). Interpreting a dynamic and uncertain world: High-level vision. *Artificial Intelligence Review, 9*(1), 37–63.

Howarth, R. J., & Buxton, H. (1992). An analogical representation of space and time. *Image and Vision Computing, 10*(7), 467–478.

Howarth, R. J., & Buxton, H. (1993). Selective attention in dynamic vision. In *Proceedings of the Thirteenth IJCAI Conference,* Chambéry, France (pp. 1579–1584). San Mateo, CA: Morgan Kaufman.

Hutchins, E. (1983). Understanding Micronesian navigation. In G. Dedre & A. L. Stevens (Eds.), *Mental models* (pp. 191–225). Hillsdale, NJ: Lawrence Erlbaum Associates.

Kalagnanam, J., Simon, H. A., & Iwasaki, Y. (1991). The mathematical bases for qualitative reasoning. *IEEE Expert: Intelligent Systems and their Applications, 6*(2), 11–19.

Koch, C., & Ullman, S. (1985). Shifts in selective visual attention: Towards the underlying neural circuitry. *Human Neurobiology, 4,* 219–227.

Kuipers, B., & Levitt, T. (1988). Navigation and mapping in large-scale space. *AI Magazine, 9*(2), 25–43.

LaBerge, D. L. (1990). Attention. *Psychological Science, 1*(3), 156–162.

Levinson, S. C. (1983). *Pragmatics.* Cambridge, England: Cambridge University Press.

Mahoney, J. V., & Ullman, S. (1988). Image chunking defining spatial building blocks for scene analysis. In Z. W. Pylyshyn (Ed.), *Computational processes in human vision: An interdisciplinary perspective* (pp. 169–209). Norwood, NJ: Ablex.

Mukerjee, A., & Joe, G. (1990). A qualitative model for space. In *Proceedings of the Eighth AAAI Conference* (pp. 721–727). Boston, MA: AAAI Press/The MIT Press.

Murray, D. W., McLauchlan, P. F., Reid, I. D., & Sharkey, P. M. (1993). Reactions to peripheral image motion using a head/eye platform. In *Proceedings of the Fourth International Conference on Computer Vision* (pp. 403–411). Berlin, Germany: IEEE Computer Society Press.

Paul, R. P. (1981). *Robot manipulators: Mathematics, programming and control.* Cambridge, MA: MIT Press.

Pylyshyn, Z. W., & Storm, R. W. (1988). Tracking multiple independent targets: Evidence for a parallel tracking mechanism. *Spatial Vision, 3*(3), 179–197.

Retz-Schmidt, G. (1988). Various views on spatial propositions. *AI Magazine, 9*(2), 95–105.

Rock, I. (1983). *The logic of perception*. Cambridge, MA: MIT Press.

Talmy, L. (1983). How language structures space. In H. L. Pick & L. P. Acredolo (Eds.), *Spatial orientation: Theory, research and application* (pp. 225–282). New York: Plenum Press.

Treisman, A. (1985). Preattentive processing in vision. *Computer Vision, Graphics and Image Processing, 31*, 156–177.

Wood, D. (1985). An isothetic view of computational geometry. In G. Toussaint (Ed.), *Computational geometry* (pp. 429–459). New York: North-Holland.

3

Time-Dependent Generation of Minimal Sets of Spatial Descriptions

Anselm Blocher
Eva Stopp
Universität des Saarlandes

Talking about a location within a geometrically represented environment implies the translation of coordinates into suitable natural language expressions. Therefore, a reference semantics anchored in the given geometry has to be developed in order to be able to describe spatial relations between objects. Answering the question "Where is object X?" will lead to a large set of spatial propositions like "X is near Y," "X is to the right of Z," and so forth. Aiming for a description of the specified location that is both easily understood and highly informative for a human being, the elements of this set must be analyzed with respect to the degrees of applicability, compatibility, uniqueness, and facility to be memorized. Furthermore, the actual context as well as presumptions about the user's knowledge must be taken into account.

Usually, in dialogue situations it is a priori unknown how much time is given for the generation of an utterance. We therefore demonstrate a first approach of applying anytime algorithms that ensure the possibility of information retrieval at any moment with increasing quality according to the amount of time available.

1. INTRODUCTION

Although considerable research has been done in each individual area of artificial intelligence (AI), the relations between seeing and speaking have only been examined in the last few years. One of the main problems that has been considered in this context is the subject of spatial descriptions and their interpretation by a human being. An overview of research in this topic can be found in Aurnague, A. Borillo, M. Borillo, and Bras (1993), Freksa and Habel (1990), Habel, Herweg, and Rehkämper (1989), Hoeppner (1990), and McKevitt (1994a, 1994b). The VITRA project (VIsual TRAnslator) as part of the German special collaboration program SFB 314—AI and Knowledge-Based Systems—has been dealing with these problems since 1985. Its aim is to develop a completely operational form of reference semantics for what is visually perceived.

If we want to be able to answer questions like "Where is object X?" then we use this reference semantics to produce a set of applicable spatial prepositions: "X is near Y," "X is between A and B," and so on. In this chapter, we describe some techniques for forming suitable prepositions as a basis for the generation of answers to "where" questions, taking into account presumptions about the user's knowledge in the actual context. Because in many dialogue situations we do not know how much time we have to find the answer, we are looking for an algorithm that can provide a result even if it has been interrupted. With increasing available time the quality of the result should increase, too.

2. THE VITRA PROJECT

In VITRA, different domains of discourse and communicative situations are considered in order to examine natural language access to visual information: In the system CITYTOUR (Schirra, Bosch, Sung, & Zimmermann, 1987) questions about observations in traffic scenes are answered, SOCCER (André, Herzog, & Rist, 1988) generates simultaneous reports of short soccer scenes, MOSES (Maaß, Wazinski, & Herzog, 1993) describes routes based on a three-dimensional model of the campus of the Saarland University, and KANTRA (Stopp, Gapp, Herzog, Längle, & Lüth, 1994) is a system for natural language access to an autonomous mobile robot. In all these domains spatial descriptions play an important role. Here, we concentrate on the SOCCER domain with respect to the search for suitable spatial descriptions.

As mentioned previously, SOCCER simultaneously analyzes and describes short soccer scenes comparable to a radio report; that is, the audience is not able to view the game itself. The input for SOCCER is the so-called *geometrical scene description*: the shape and configuration of the soccer field as the *static background*, together with the positions and velocity vectors of every mobile object perceived in the soccer field as *mobile object data*. The latter is generated by the motion analysis system ACTIONS (Herzog et al., 1989) and is delivered for every time quantum.

The core system of SOCCER consists of three components (cf. Fig. 3.1): The incoming geometrical scene description is analyzed by the Event Recognition component. Here, the given percepts are interpreted as

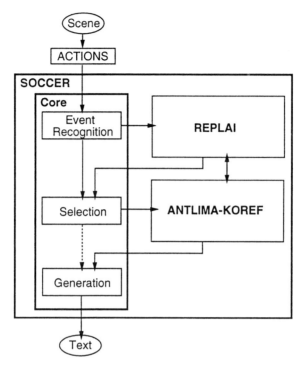

FIG. 3.1. Extended architecture of SOCCER.

instances of spatial and spatio-temporal relations like *in front of* or *running-pass*. Spatio-temporal relations are also called *events*. The set of recognized events is given to the Selection component, which then has to choose one of them to continue the running report. The selected event is then transformed by the Generation component into a German utterance, taking into account the coherence of the whole report. This core system has been extended by two components: REPLAI (Retz-Schmidt, 1992), which recognizes the plans and intentions of the observed agents, and ANTLIMA-KOREF (cf. Blocher, 1994; Blocher & Schirra, 1995; Schirra, 1994; Schirra & Stopp, 1993), a listener model with mental images.

3. SPATIAL RELATIONS IN VITRA

Following Herskovits (1986), we distinguish between the "basic meaning" of a preposition itself and an instance of it with respect to some arguments: an *object to be localized* (LO) and one or more *reference objects* (RO).[1] This so-called spatial proposition is represented as follows:

---

[1]Because people often do not account for details we can also rely on idealizations of the concerned objects (cf. Gapp, 1994a).

(< relation name > < LO > < sequence of ROs >)

In the sentence *"The child is near the house,"* the basic meaning of the spatial concept near is an abstraction from the concrete situation, and thus only dependent on a distance parameter relative to the concerned objects. To encode these *essential parameters* we use cubic spline functions with values between 0 and 1 (1 stands for absolutely applicable, 0 for not at all applicable). These functions are adjustable to ongoing empirical experiments (cf. Gapp, 1995). In combination with RO-dependent *instantiation rules*, for every instance of a spatial concept a proposition is created to which a *degree of applicability* can be applied with respect to the position of LO (cf. Gapp, 1994a; Stopp et al., 1994).

We distinguish between three classes of static spatial relations that differ in the types of parameters according to their basic meaning:

- *Topological relations:* These relations are only dependent on the so-called *local distance*, that is, the distance between LO and RO, scaled by the RO's extension (e.g., *at, near*).
- *Projective relations:* The relations *in front of, behind, left, right, below, above,* and *beside* primarily depend on the *scaled local deviation angle* from their canonical direction. In most cases, this parameter is competent for the computation of the degree of applicability of such a relation. In some cases, the local distance may also play a role in the interpretation of the relation.
- *Between:* This relation takes an exceptional position: In VITRA it is handled as a combination of LO *being in front of RO-1, as seen from RO-2* and vice versa.

This approach for the evaluation of spatial relations can be used for both two- and three-dimensional domains. Figure 3.2 shows the applicability region of an object being *in front of* the "Left-Goal" in *intrinsic* use; that is, the canonical direction is implied by the prominent front of the goal. Following Retz-Schmidt (1988), we also distinguish two other forms of canonical direction: *extrinsic*, which means "seen from another object," and *deictic* as a special case of extrinsic use: "seen from a known viewpoint."

4. ANYTIME ALGORITHMS

So far, research in artificial intelligence concerning the connection between visual and verbal space has been done under the assumption that the delivered visual information was always complete and unlimited re-

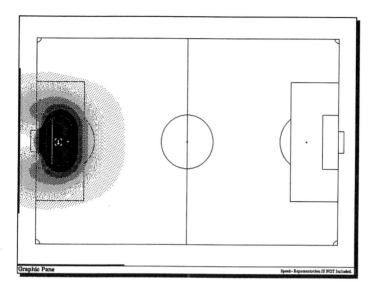

FIG. 3.2.   Graphical representation of the classification functions for being *in front of* the left goal, including a distance concept (darker parts of the region correspond to higher applicability values).

sources were available for the computation. If we want to treat visual information in a more realistic way we must adjust the increasing complexity by intelligently applying existent resources. This is necessary because in some situations, for example, in dynamic environments like the field of route descriptions, it is important to guarantee an answer even if there is not enough time to find the perfect one. Russell and Wefald (1991) identified three kinds of time-dependent tasks:

- A fixed time limit is given before starting the algorithm, after which a result should be delivered.
- A fixed cost function allows the computation of the time available.
- The limit is uncertain. Here, in most cases a probability distribution is used in order to estimate the time available.

Anytime algorithms can also be considered as a special case of interruptable algorithms (cf. Dean & Boddy, 1988) with the following properties: At any time, they can be interrupted and restarted; their results can be seen as a monotonically improving function of the invested computing time with respect to a measure of quality. In contrast to applying anytime algorithms to iterative approximation procedures, the interruptability of complex cognitive processes, such as natural language processing, planning, deduction, and visual perception, has barely been examined yet. In these complex symbolic tasks, the concept of transactions limits the anytime behavior (cf. Görz, 1994).[2] In order to assess the quality of the produced result of an anytime algorithm, *performance profiles* that improve monotonically $Q_A(t)$ (cf. Russell & Wefald, 1991) are defined dependent on the running time t of algorithm A.[3]

5. BUILDING SPATIAL DESCRIPTIONS

The task of localizing an object LO to answer questions like "Where is the ball?" can be divided into two parts: the search for reference objects and the evaluation of spatial relations. The complexity of building propositions requires the application of criteria to reduce the large number of possible relations and reference objects. Even in a quite small domain like SOCCER with nearly 150 static or movable objects and about 60 spatial relations this leads to a combinatoric explosion, especially considering three-place propositions like *(rel LO RO-1 RO-2)*. The set of possible reference objects can be restricted by applying the following criteria:

- *Distance:* Objects being closer to LO than others should be preferred for selection as ROs.
- *Salience:* Objects more remarkable in color, size, shape, and so on, than others should be preferred.
- *Linguistic context:* Prefer objects that have already been mentioned and that are still in the focus of the communication partner.

In employing anytime algorithms, we have to guarantee two aspects. First, there must always be an applicable result available—even at the very start of the algorithm. Second, with an increasing running time, which corresponds with an increasing number of relations and reference objects tested, the result has to be improved according to a monotonous performance profile. Additionally, we integrate a third aspect: The strongest improvement of the resulting locative description should take place at the beginning of the process (cf. Fig. 3.3). Therefore, we build up a qualitative hierarchy of sets of reference objects: First, we consider those ROs fulfilling the strongest restriction, followed by the group of ROs

---

[2]A transaction is a sequence of operations leading from one consistent state to another without any possibility of interruption. As an example, Görz (1994) described an anytime chart parser based on feature unification where the unification transactions themselves cannot be interrupted.

[3]Menzel (1994) introduced the notion of weak anytime algorithms that also have monotonical performance profiles without being identical for every application: He described a constraint-based and resource-adaptive parsing algorithm.

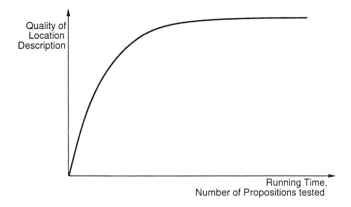

FIG. 3.3. Sketch of the performance profile of the implemented algorithm.

satisfying a weaker one, and so forth. If sufficient time is available, *all* reference objects can be tested. To construct the aforementioned hierarchy, we chose in our approach a kind of variable distance criterion dependent on the physical and referential structure of the environment. In a first step we look for the smallest two- or three-dimensional region R satisfying the proposition *(in LO R)*, that is, that this proposition is fully applicable. Using both geometrical knowledge about the constituents of R and conceptual domain knowledge we then build up a set of possible reference objects. The following example (cf. Fig. 3.4) in the SOCCER domain illustrates the construction by recursion of the so-called *In-Path* leading to R:

Per definitionem the object "World" describes the convex hull of all domain objects. Therefore, the proposition *(in LO World)* is always applicable and we can set the initial value of R to "World." This guarantees the aforementioned first aspect of anytime algorithms: being able to deliver an applicable result from the very beginning. Using conceptual knowledge about the object "World," we now consider all its direct constituent regions.[4] In this first approach without loss of generality we assert that these regions will not overlap. Adding a—perhaps empty—*Complement* region to cover those parts that have not been specified explicitly, provides a perfect decomposition of "World."[5]

In our example, we obtain three defined subregions of "World": "Left Goal," "Field," "Right Goal," and the *Complement* region, which in this case could be described as *"Being neither in the left or right goal nor in the field."* We now explore the defined subregions: If for a subregion SR the proposition *(in LO SR)* is applicable, R will be replaced by SR: here, "Field"

---

[4]We do not need to cope with, for example, points or lines, as the spatial relation *in* can only be applied to two- or three-dimensional regions.

[5]In the following, we do not mention the *Complement* region if it is empty.

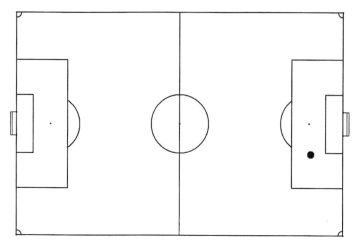

FIG. 3.4. A position of LO.

replaces "World." Note that the spatial relation *in* can at most hold for one of the subregions, because they do not overlap. A new pass of the algorithm starts working on the subregions of "Field." Finally, with R being the "Right Penalty Area," *in* cannot be applied to the subregions of R (in fact, there is only one, the "Right Goal Area"). Obviously LO is positioned in the *Complement*; that is, R remains unchanged and the algorithm of constructing the *In-Path* terminates. The *In-Path* can be represented as a tree or a queue, with *in* applied to R at its top (cf. Fig. 3.5).

The region R found in the preceding paragraph represents the core of the set of possible reference objects. This set consists of the following elements: all the geometrical constituents of R as, for example, lines, points, and so on, as well as objects specified as conceptual parts of R: For instance, if LO is *in* the "Right Penalty Area," both the "Right Goal Area" and the "Right Penalty Spot" become ROs. Additionally, all those dynamic objects being *in* R have to be taken into account.

Concerning the selection of spatial relations, we propose considering "simple" binary relations first. The following shows the default series of sorted spatial relations, which can be changed on user demand by adding the desired set to the top-level function of the algorithm.[6]

- *Topological relations:* Here, we do not use relations like *away from* that result in a kind of negative description limiting the region of *not* being applicable.

---

[6]For simplification, in our examples we use a two-dimensional representation of the Soccer domain from a bird's-eye view.

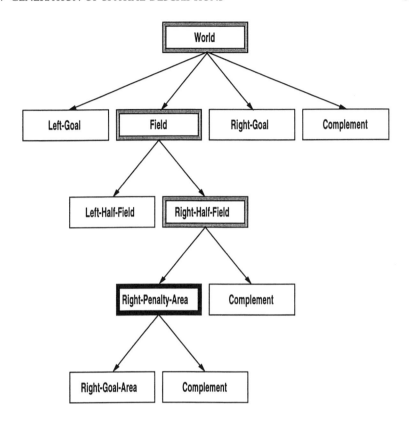

In-Path(LO) = ((in LO Right-Penalty-Area)
              (in LO Right-Half-Field)
              (in LO Field)
              (in LO World))

FIG. 3.5. The corresponding *In-Path* to Fig. 3.4.

   −*kontakt* (at, being in contact with)
   −*bei* (close to)
   −*nahe* (near)
• *Projective relations in intrinsic and/or deictic use:* For the task of describing object locations, it is preferable to employ projective relations with an integrated distance concept.
   −*links* (to the left of)
   −*rechts* (to the right of)
   −*vor* (in front of)
   −*hinter* (behind)

- *Zwischen (between):* This relation has already been described; note that it results in a more complex three-place proposition.

Finally, extrinsic use of projective relations is tested. As these propositions need a second RO as an additional argument—which in contrast to deictic use is not known already—their calculation is more expensive and therefore put at the end of the process. This guarantees better performance in the beginning; that is, a highly acceptable proposition will usually be found very quickly, whereas looking for further improvement will take longer.

In Table 3.1 an exemplary output demonstrates the improvement of the results of the location description according to successive queries; the LO

TABLE 3.1
An Exemplary Output

HP-LUCID:ANT: (anywhere (list referee): views '(intrinsic))
Start Time: 14:04.31
Query Time: 14:04.31
    ((IN REFEREE WORLD) 1.0)
Query Time: 14:04.33
    ((NAHE REFEREE FRONT-LINE-RIGHT-PENALTY-AREA) 0.95)
    ((IN REFEREE RIGHT-PENALTY-AREA) 1.0)
    ((IN REFEREE RIGHT-HALF-FIELD) 1.0)
    ((IN REFEREE FIELD) 1.0)
    ((IN REFEREE WORLD) 1.0)
Query Time: 14:04.34
    ((RECHTS REFEREE RIGHT-GOAL-AREA) 0.98)
    ((NAHE REFEREE RIGHT-GOAL-AREA) 0.98)
    ((IN REFEREE RIGHT-PENALTY-AREA) 1.0)
    ((IN REFEREE RIGHT-HALF-FIELD) 1.0)
    ((IN REFEREE FIELD) 1.0)
    ((IN REFEREE WORLD) 1.0)
Query Time: 14:04.36
    ((RECHTS REFEREE RIGHT-GOAL-AREA) 0.98)
    ((NAHE REFEREE RIGHT-GOAL-AREA) 0.98)
    ((IN REFEREE RIGHT-PENALTY-AREA) 1.0)
    ((IN REFEREE RIGHT-HALF-FIELD) 1.0)
    ((IN REFEREE FIELD) 1.0)
    ((IN REFEREE WORLD) 1.0)
Query Time: 14:04.38
    ((ZWISCHEN REFEREE RIGHT-SIDE-LINE-RIGHT-PENALTY-AREA RIGHT-
PENALTY-SPOT) 1.0)
    ((RECHTS REFEREE RIGHT-GOAL-AREA) 0.98)
    ((NAHE REFEREE RIGHT-GOAL-AREA) 0.98)
    ((IN REFEREE RIGHT-PENALTY-AREA) 1.0)
    ((IN REFEREE RIGHT-HALF-FIELD) 1.0)
    ((IN REFEREE FIELD) 1.0)
    ((IN REFEREE WORLD) 1.0)
HP-LUCID:ANT:

"Referee" is positioned as shown in Fig. 3.4.[7] As a preselection of all applicable propositions found during the process, only the best result of each group of relations (topological, projective, and *between*), as well as the complete *In-Path*, is returned. The final spatial description should not combine more than two propositions (topological/projective or projective/projective) with identical RO (cf. Gapp, 1994b)—*between* being already a combination—and additionally mention the *in*-relation, if necessary.

The *In-Path* can be used to enlarge the set of possible ROs by searching in the next largest *in*-region—in the current example the "Right Half Field." This would be useful if the set of propositions found in the first pass had low degrees of applicability. Furthermore, a lack of understanding of the user can be clarified by dialogues as for instance: *U*: "OK, the referee stands in the right penalty area, but where is the right penalty area?"—*S*: "In the right half field."

6. THE ROLE OF UNDERSPECIFIC DEFINITE DESCRIPTIONS

A localization expression as an answer to a "where" question should not only be correct but also unique, easy to memorize, and efficient to decode for a listener.[8] In order to produce a description that fulfills these properties, we must consider the listener's knowledge about the context. For this task, we use ANTLIMA, a listener model with mental images (cf. Blocher, 1994; Blocher & Schirra, 1995; Blocher, Stopp, & Weis, 1992; Schirra, 1994): Instances of spatial relations are verified with respect to the typicality of their appearance in the context. Because the same methods are used as for the generation of spatial relations, the so-called *typicality value* (t-value) corresponds to the degree of applicability.

For any given object, there is a minimal unique definite description (or even several of them), which usually consists of a set of attributes that separates it from all other objects in the context. If this set is rather large, shorter descriptions become ambiguous. But, if the concerned object is focused on by the listener these *underspecific definite descriptions* (UDDs) could nevertheless be understood within the context. In most cases, this is used for objects that have already been mentioned in an earlier utterance (cf. Jameson & Wahlster, 1982). As we do not use a text memory—in contrast to, for example, the techniques introduced in André et al. (1988)— we have no *explicit* knowledge of what is in the user's focus. Instead,

---

[7]For simple reasons of space we only applied intrinsic use to projective relations.

[8]This corresponds to the conversational maxims of Grice that the listener should be informed about all relevant facts avoiding redundancy (cf. Grice, 1975).

ANTLIMA provides the visual context *implicitly* in the form of the geometrical model of the listener's mental image. Accordingly, we can use spatial relations to retrieve information about the presumed user knowledge.

Underspecific definite descriptions can improve a locative expression with respect to its conciseness in two ways: by either denoting a part of an object by naming the whole object or discarding some specifications implicitly known to the listener. An example for the former would be the replacement of the object "Upper Sideline" by its overall object "Outline." The description remains understandable if the listener does not focus on another part of the "Outline," which in the SOCCER domain is composed of two sidelines and two endlines. The latter, discarding implicitly redundant specifications, is useful, if an object happens to occur in more than one instantiation as, for example, corners, penalty areas, goals, and so on. If we know that the object "Upper Left Corner" is preferred among all other instances of the kind "Corner" that are in the listener's mental focus with respect to the current context, the explicit mentioning of this attribute is obsolete. The use of *(near Referee Corner)* is sufficient for describing the location of "Referee" whereas further identification is done by the listener. The generation of these ellipses is called the *elision of adjectives and/or attributes* (cf. Jameson & Wahlster, 1982). Table 3.2 shows the actually implemented UDDs and the elements subsumed by them.

If we have a proposition with an UDD, for example, *(near Referee Corner)* (cf. Fig. 3.6), a corresponding proposition is created for every element of the set M(UDD), which is subsumed under the notion UDD, here:

- (near Referee Upper Left Corner)   • (near Referee Upper Right Corner)
- (near Referee Lower Left Corner)   • (near Referee Lower Right Corner)

TABLE 3.2
Underspecific Definite Descriptions

Goal	→	(Left ∨ Right) Goal
Goal-Post_Left_Goal	→	(Left ∨ Right) Goal-Post_Left_Goal
Goal-Post_Right_Goal	→	(Left ∨ Right) Goal-Post_Right_Goal
Goal-Post	→	Goal-Post (Left_Goal ∨ Right_Goal)
Goal-Area	→	(Left ∨ Right) Goal-Area
Penalty-Area	→	(Left ∨ Right) Penalty-Area
Penalty-Spot	→	(Left ∨ Right) Penalty-Spot
Left_Corner	→	(Upper Left ∨ Left Lower) Corner
Right_Corner	→	(Upper Right ∨ Lower Right) Corner
Lower_Corner	→	(Lower Left ∨ Lower Right) Corner
Upper_Corner	→	(Upper Left ∨ Upper Right) Corner
Corner	→	(Left ∨ Right) Corner
Endline	→	(Left ∨ Right) Endline
Sideline	→	(Upper ∨ Lower) Sideline
Outline	→	Endline ∨ Sideline

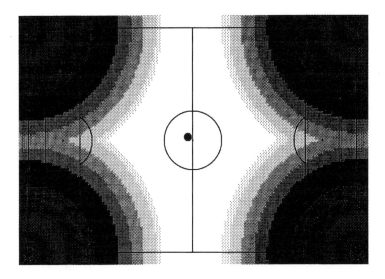

FIG. 3.6.   The degrees of applicability for *(near Referee Corner)*.

The typicality value of UDD results from the maximum of the t-values of these propositions:

$$T\text{-}Value(UDD) = max_{RO \in M(UDD)}T\text{-}Value(RO)$$

So, the usage of underspecific definite descriptions allows the processing of ambiguous propositions in the mental model. We profit from this in our considerations about the possible elision of attributes of a reference object by replacing the RO of a detected proposition with the corresponding UDD and computing the t-value of this new proposition. If the typicality values of RO and UDD are the same, we can erase the specification that was left out during the transition from RO to UDD because the listener is able to infer it. This process is continued with the new RO until no superior UDD is found or the typicality values of RO and UDD differ (cf. Fig. 3.7). Thus, the speaker can use, for example, the expression "Corner" instead of "Upper Left Corner" without loss of understanding, even if the object meant itself was never explicitly referred to before. If only this particular corner is the preferred one in the imaginative visual focus of the listener, the attributes "upper" and "left" can be elided.

7. CONCLUSION

In this chapter, we have presented an approach for describing localizations of objects in their actual environment, especially for answering "where" questions. This task can be divided into two parts: the search for reference

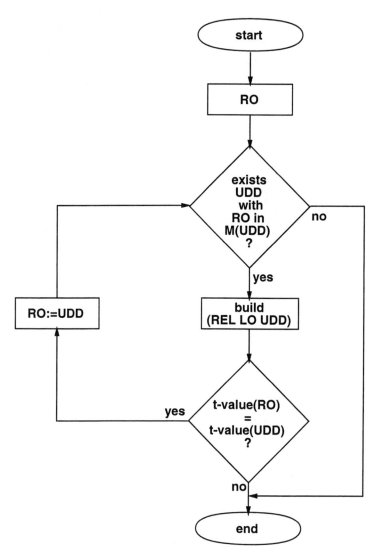

FIG. 3.7. The construction of underspecific definite descriptions.

objects and the evaluation of spatial relations. Considering the combinatoric explosion in constructing spatial propositions we looked for a possibility of restricting the complexity of the algorithm, especially
concerning the number of tested instances.

Because in some domains, for example, for route descriptions, algorithms should be interruptable, we introduced a first approach of an

anytime algorithm for this task of locative descriptions. Being able to deliver an applicable result directly from the start, the algorithm shows a performance profile that improves most rapidly at the beginning and flattens out toward the end.

In order to be able to discard redundant information we use the imaginative visual focus of the listener to find out underspecific definite descriptions that can still be understood.

The implementation of the algorithm has been done in LISP with CLOS in an object-oriented manner.

REFERENCES

André, E., Herzog, G., & Rist, T. (1988). On the simultaneous interpretation of real world image sequences and their natural language description: The system SOCCER. In *Proceedings of the 8th ECAI* (pp. 449–454). London: Pitman.

Aurnague, M., Borillo, A., Borillo, M., & Bras, M. (Eds.). (1993). *Proceedings of the 4th International Workshop on Semantics of Time, Space and Movement and Spatio-Temporal Reasoning*. Toulouse, France: Groupe "Langue, Raisonnement, Calcul".

Blocher, A. (1994). *KOREF: Zum Vergleich intendierter und imaginierter Äußerungsgehalte* [*KOREF*: Comparing the intended and anticipated understanding]. (Memo 61, Project VITRA). Saarbrücken, Germany: Saarland University, Department of Computer Science.

Blocher, A., & Schirra, J. R. J. (1995). Optional deep case filling and focus control with mental images: ANTLIMA-KOREF. In *Proceedings of the 14th IJCAI* (pp. 417–423). San Mateo: CA: Morgan Kaufmann.

Blocher, A., Stopp, E., & Weis, T. (1992). *ANTLIMA-1: Ein System zur Generierung von Bildvorstellungen ausgehend von Propositionen* [ANTLIMA-1: Generating mental images from propositions]. (Memo 50, Project VITRA). Saarbrücken, Germany: Saarland University, Department of Computer Science.

Dean, T., & Boddy, M. (1988). An analysis of time-dependent planning. In *Proceedings of AAAI-88* (pp. 49–54). Minneapolis, MN: Morgan Kaufmann.

Freksa, C., & Habel, C. (Eds.). (1990). *Repräsentation und Verarbeitung räumlichen Wissens* [Representation and processing of spatial knowledge]. Berlin: Springer.

Gapp, K.-P. (1994a). Basic meanings of spatial relations: Computation and evaluation in 3d space. In *Proceedings of AAAI-94* (pp. 1393–1398). Menlo Park, CA: AAAI Press.

Gapp, K.-P. (1994b). A computational model of the basic meanings of graded composite spatial relations in 3d space. In *Proceedings of the AGDM'94 Workshop* (pp. 66–79). Delft: Netherlands Geodetic Commission.

Gapp, K.-P. (1995). An empirically validated model for computing spatial relations. In I. Wachsmuth, C.-R. Rollinger, & W. Brauer (Eds.),. *KI-95: Advances in artificial intelligence. 19th annual German conference on artificial intelligence* (pp. 245–256). Berlin: Springer.

Görz, G. (1994). Anytime algorithms for speech parsing? In *Proceedings of Coling-94* (pp. 997–1001). Kyoto, Japan: Association of Computational Linguistics.

Grice, H. P. (1975). Logic and conversation. In P. Cole & J. L. Morgan (Eds.), *Syntax and semantics* (pp. 41–58). New York: Academic Press.

Habel, C., Herweg, M., & Rehkämper, K. (Eds.). (1989). *Raumkonzepte in Verstehensprozessen: Interdisziplinäre Beiträge zu Sprache und Raum* [Spatial concepts in comprehension processes: Interdisciplinary contributions to language and space]. Tübingen, Germany: Niemeyer.

Herskovits, A. (1986). *Language and spatial cognition. An interdisciplinary study of the prepositions in English.* Cambridge, England: Cambridge University Press.

Herzog, G., Sung, C.-K., André, E., Enkelmann, W., Nagel, H.-H., Rist, T., Wahlster, W., & Zimmermann, G. (1989). Incremental natural language description of dynamic imagery. In C. Freksa & W. Brauer (Eds.), *Wissensbasierte Systeme. 3. Int. GI-Kongreß* (pp. 153–162). Berlin: Springer.

Hoeppner, W. (Ed.). (1990). *Workshop: Räumliche Alltagsumgebungen des Menschen* [Spatial everyday environments of human beings]. Koblenz, Germany: University of Koblenz.

Jameson, A., & Wahlster, W. (1982). User modelling in anaphora generation: Ellipsis and definite description. In *Proceedings of the 5th ECAI* (pp. 222–227). Orsay, France.

Maaß, W., Wazinski, P., & Herzog, G. (1993). VITRA GUIDE: Multimodal route descriptions for computer assisted vehicle navigation. In *Proceedings of the Sixth International Conference on Industrial and Engineering Applications of Artificial Intelligence and Expert Systems IEA/AIE-93* (pp. 144–147). Edinburgh, Scotland.

McKevitt, P. (Ed.). (1994a). The integration of natural language and vision processing [Special issue]. *Artificial Intelligence Review journal, 8.*

McKevitt, P. (Ed.). (1994b). *Proceedings of AAAI-94 Workshop on "Integration of Natural Language and Vision Processing."* Seattle, WA: AAAI Press.

Menzel, W. (1994). Parsing natural language under time constraints. *Proceedings of the 11th ECAI* (pp. 560–564). New York: Wiley.

Retz-Schmidt, G. (1988). Various views on spatial prepositions. *AI Magazine, 9,* 95–105.

Retz-Schmidt, G. (1992). *Die Interpretation des Verhaltens mehrerer Akteure in Szenenfolgen* [Interpreting the behavior of several actors in scene sequences]. Berlin: Springer.

Russell, S., & Wefald, E. (1991). *Do the right thing: Studies in limited rationality.* Cambridge, MA: MIT Press.

Schirra, J. R. J. (1994). *Bildbeschreibung als Verbindung von visuellem und sprachlichem Raum: Eine interdisziplinäre Untersuchung von Bildvorstellungen in einem Hörermodell* [Image description as a connection of visual and verbal space: An interdisciplinary study of mental images in a listener model]. St. Augustin, Germany: infix.

Schirra, J. R. J., Bosch, G., Sung, C.-K., & Zimmermann, G. (1987). From image sequences to natural language: A first step towards automatic perception and description of motions. *Applied Artificial Intelligence, 1,* 287–305.

Schirra, J. R. J., & Stopp, E. (1993). ANTLIMA—A listener model with mental images. In *Proceedings of the 13th IJCAI* (pp. 175–180). San Mateo, CA: Morgan Kaufmann.

Stopp, E., Gapp, K.-P., Herzog, G., Längle, T., & Lüth, T. C. (1994). Utilizing spatial relations for natural language access to an autonomous mobile robot. In B. Nebel & L. Dreschler-Fischer (Eds.), *KI-94: Advances in artificial intelligence* (pp. 39–50). Berlin: Springer.

4

A Computational Model
for the Interpretation
of Static Locative Expressions

Vittorio Di Tomaso
Scuola Normale Superiore, Pisa

Vincenzo Lombardo
Leonardo Lesmo
Università di Torino

This chapter presents an approach to the interpretation of prepositions in locative expressions, that is, expressions such as *the cat is on the mat*. Such expressions are used to describe a scene where an object is located, or is seen as in motion along a path, with respect to a reference object. The meaning of a preposition is viewed as a set of *ideal meanings*, each of which represents a *spatial relation*, which can be specialized on the basis of its actual arguments. Ideal meanings are simple geometric relations that can be sense-shifted, consequently defining a taxonomy of relations. The chapter describes how the interpretation process disambiguates and specializes the meaning of the preposition given a fragment of spatial knowledge on objects and events.

1. INTRODUCTION

This chapter is concerned with the representation of the meaning of locative prepositions and with the use of that representation in the analysis of natural language sentences. In particular:

- The representation should distinguish between homonymy and polysemy. In other words, each preposition must be allowed to have

73

different independent meanings and, for each of them, various
degrees of indeterminacy and specificity.

• The representation should enable the interpreter to disambiguate the
preposition, that is, to select a given meaning and to reduce its
indeterminacy, by specializing the sense of the preposition on the
basis of the context where the preposition occurs.

The disambiguation problem cannot be solved unless detailed information about the objects involved in the scene described by the locative
expression is available. The chapter claims that a strict interaction is
needed between the representation of spatial prepositions and the general
world knowledge encoded in the dictionary/encyclopedia.[1]

In particular, the required information concerns the spatial properties
of the entities involved in the description. We argue that such information
is part of the conceptual representation of physical objects[2]: The fact that
an entity is flat, or that it can be conceptualized as a container, or that it
has a cognitively salient surface (as a table), must be available to the
interpreter, as the following examples show:

(1) a. Il ragno è sul muro
         The spider is on the wall
     b. Il ragno è sul tavolo
         The spider is on the table
     c. ??La scatola è sul muro
         The box is on the wall

It seems reasonable to assume that in (1) a single relation, but variously
specified, relates the objects in each sentence. It is only the knowledge
about the verticality of walls and the climbing ability of spiders that
enables a reader to understand that (or to imagine a situation where) the
spider is attached to the wall, as in (1a). A different spatial situation is
imagined in (1b), and the same knowledge accounts for the very dubious
acceptability of (1c). It must be observed that the strangeness of (1c) does
not depend on the difficulty to imagine a proper situation; in fact "The
box is attached to the wall" is easily understandable. On the contrary, it

---

[1]It is at least since Quine (1951) that philosophers of language questioned such distinction.
In fact, the point of the dictionary/encyclopedia distinction is the sorting out of information
into semantically relevant and semantically irrelevant, that is, into information that
contributes to a word's meaning and information that does not, but unfortunately it turned
out that it is very difficult (if not impossible) to decide where such a boundary is (see
Marconi, 1997, for discussion).

[2]The chapter does not address metaphorical uses of locative expressions such as *I have
an idea in my head.*

seems that in (1c) there is a problem in the processing of the expression, due to the difficulty to specialize the meaning of the preposition on the basis of the available information: "su" *(on)* cannot be properly specialized on the basis of the actual pair of objects <a box, a wall>.

It is interesting to note that the specialization process presents some analogies with type coercion in the manner of Pustejovsky (1991). Also in the case of prepositions, in fact, the knowledge about the arguments has a fundamental impact on the interpretation of a relation (expressed via a preposition and not via a verb). The main difference is that we do not assume any fixed structure in the representation of objects (such as Qualia Structure proposed by Pustejovsky): Spatial properties are represented together with all other properties of a given object.

In this chapter we deal exclusively with static spatial locative expressions, that is, expressions such as the ones appearing in (1), where the location of an object is described via a reference to the known location of another object. In the literature, the first object is called the *relatum* (Miller & Johnson-Laird, 1976), the *figure* (Talmy, 1985), or the *trajector* (Lakoff, 1987; Langacker, 1991) and the second object is respectively called the *referent*, the *ground*, or the *landmark*. We follow Herskovits (1986) and use the terms *located object* and *reference object*.

The following section includes a short review of the current theories on the meaning of prepositions. The next three sections describe our approach: lexical and terminological knowledge, semantic representation, and interpretation. An example and some conclusions close the chapter.

2. THEORIES OF THE MEANING OF PREPOSITIONS AND ON SPATIAL RELATIONS

Theories of the meaning of locative prepositions can be classified as *simple relations models* versus *multiple relations models* (cf. Herskovits, 1986). The first class includes theories coming both from linguistics (Bennett, 1975) and from the computational paradigm (Boggess, 1979; Cooper, 1968; Miller & Johnson-Laird, 1976; Waltz, 1980). Presumably, the position of Jackendoff (1983, 1990) also falls in this class. The basic idea of these approaches is that all spatial uses of a given preposition can be derived from a single geometric relation, whose arguments are the reference (background) object and the located object. This position can be summarized by the following claim:

the meaning of a preposition in an expression NP + PP[Prep + NP] is a relation $R(x,y)$, where x is the located object, y is the reference object, and R is a spatial relation that gives the information needed to narrow the domain of search of the located object x (cf. Johnson-Laird, 1983, pp. 201–203 and pp. 250–265; Miller & Johnson-Laird, 1976, p. 379),

where spatial relations are defined independently of their arguments. Although the basic statement is reasonable, the attempt to apply the approach in an actual interpretation of prepositions reveals some draw-backs: (a) disjunctive definitions of prepositional relations, which obscures the polysemy of prepositions, without devising criteria for disam-biguation; (b) difficulty of accounting for all spatial inferences (that can also depend on further factors, like relevance of certain objects in the human communication and in the interaction with the physical world); (c) impossibility of explaining cases where the use of a preposition di-verges from the standard geometric relation (tolerance); and (d) no ac-count for conventional deviations from the standard meaning (sense shifts).

The class of multiple relations models, exemplified by Lakoff (1987) and Herskovits (1986) and, from a computational point of view, by Japkowicz and Wiebe (1991), Kalita and Badler (1991), and Lang (1993), seems to overcome many of the objections outlined earlier, thus account-ing for the flexibility and adaptability in the use of spatial relations. The main claims are the following:

1. Each preposition defines a family of senses such that there is a "central" or "ideal" meaning and the others are obtained via *moti-vated* transformations.[3]

2. The meaning of a preposition in an expression NP + PP[Prep + NP] is a relation $R_i(x,y)$ (where x is the located object and y is the reference object) that belongs to the family of senses defined by the preposi-tion; the choice of a R_i depends on the nature of the objects x and y.

For example, Herskovits (1986) associated each preposition with one or more *ideal meanings*, which are elementary geometrical-topological relations (like inclusion or coincidence), and a number of sense shifts. Ideal meanings abstract spatial objects to geometric entities (points, lines, surfaces, volumes). Sense shifts are variations of the ideal meanings mo-tivated by particular features of the objects in relation, mostly depending

---

[3]The term *motivated* is used here in the sense of Lakoff (1987): The sense shifts of the central meaning of a preposition are due to the action of some criteria, which are not predictive principles, but can be somehow identified. Transformations can be seen as "categorizing relationships of schematicity" (Langacker, 1988).

on the geometric conceptualization of the objects or on pragmatic principles like perceptive salience and communicative relevance. Further distinctions can be drawn within the latter class. According to Vandeloise (1994), different analyses can be classified on the basis of the conceptual tools they exploit, so that we can distinguish between geometrical analyses (referring to the dimensionality of objects: Hawkins, 1988), topological analyses (referring to ideas such as inclusion: Herweg, 1989), or functional analyses (referring to the contrast between the function of the reference object and the function of the located object, e.g., contrasts like container/contained, bearer/burden: Coventry, chap. 16 of this volume; Vandeloise, 1994). Theories can also be classified with respect to the number of levels assumed to represent the uses of prepositions (multiple levels, like most of the analyses cited before, vs. a single level, like Vandeloise, 1994).

3. REPRESENTATION OF LEXICAL MEANINGS

As the basic formalism for the representation of lexical meanings we adopted an extended version of a KL-ONE-like terminological net (cf. Brachman & Schmolze, 1985).

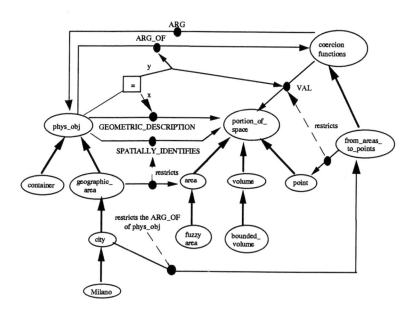

FIG. 4.1. Use of GEOMETRIC_DESCRIPTION in spatial entities representation.

Physical Objects (see Fig. 4.1) are related to the portion of space they typically identify. The region is approximated to a geometric entity, that can be mono-, bi-, or tridimensional and some topological properties are given (i.e., if it is a bounded or unbounded region). Such information is associated with the role SPATIALLY_IDENTIFIES. Moreover, alternative geometric conceptualization are specified via coercion functions. For example, *Milano* spatially identifies an *area* ("Passeggio per Milano" *I'm walking around in Milan*), but can be conceptualized as (i.e., coerced to) a point ("Vado a Milano" *I'm going to Milan*).

The coercion is applied when the PP is interpreted (see section 5) and results from the attempt to satisfy the selectional constraints imposed by the semantics of the preposition. In Fig. 4.1, the node *phys_obj* has a role SPATIALLY_IDENTIFIES, whose value restriction is a *portion of_space*, and a role GEOMETRIC_DESCRIPTION with the same value restriction. The first of them identifies the "typical" region identified by a given object, whereas the second allows the interpreter to force an alternative reading. For instance, a *city* inherits from *geographic_area* the restriction that it SPATIALLY_IDENTIFIES an area, but it can also be the argument (ARG_OF) of the function *from_areas_to_points*, which has a *point* as value (VAL), thus enabling the interpreter to coerce the standard spatial interpretation (*area*) to a new one (*point*).

Using such a representation of spatial knowledge it is possible to account for the fact that information on geometrical features of objects (or, to be more precise, of the spatial regions identified by objects) plays a crucial role in the interpretation of the preposition and to explain how the geometrical nature of an object can be coerced in order to satisfy the selectional restrictions of the preposition.

The lexical entries of prepositions refer to sets of relations in the terminological net. Each of these relations corresponds to a different ideal meaning of a preposition and the association of more than one ideal meaning with a preposition represents homonymy. For example, the preposition *in* (*in, into*) refers to the relations "located in" (*loc_in*) and "destination inward a reference object" (*dest_in*); the preposition *a* (*at, in, to*) refers to the relations "adjacency" (*loc_at*) and "destination" (*dest_to*). Of course, also meanings belonging to different cognitive domains can be distinguished in a similar way (e.g., to keep apart the "temporal" meaning of "in" from its spatial meaning). However, relations belonging to the same domain share major argument restrictions, accounting for the fact that the relations of the same domain (e.g., locative) seem intuitively closer than relations belonging to different domains (e.g., locative and temporal).

In Fig. 4.2, one of the ideal meanings of *in* (*loc_in*) is reported. Each relation has two roles (arguments), XARG and YARG, whose type restrictions (filler's type) characterize located objects and reference objects re-

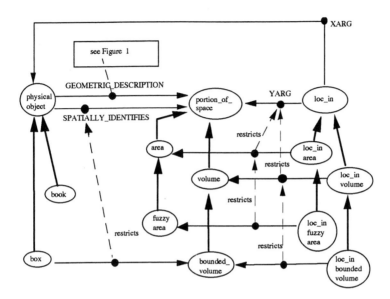

FIG. 4.2. A portion of the terminological net describing the ideal meaning *loc_in*.

spectively.[4] More specific relations, defined via restrictions on the arguments, represent the possible sense shifts. Because of the hierarchical organization of prepositional senses (cf. the lattice of image schemata of Lakoff, 1987), it is possible to specify as meaning of a preposition the top-level nodes of the hierarchy (the ideal meanings), leaving to the interpreter the task of finding the specialized relation. To summarize, homonymy of prepositions is represented by allowing several links to ideal meanings in a lexical entry; polysemy (viewed as *lack of specificity*; Zwicky & Saddock, 1975) is reflected by the hierarchy itself.

By observing the portion of terminological net shown in Fig. 4.2, it can be noted that a static locative relation takes as arguments a *physical object* and a *portion of space* (identified by a physical object). Contrary to Herskovits' (1986) approach, where two portions of space are related, we claim that the located object is usually the object itself. In our opinion it seems counterintuitive to assume an immediate extension to dynamic relations that forces an interpretation of *the spider went into the box* as "the portion of space identified by the spider moved so that at the end it was contained in the portion of space identified by the box."

---

[4]Role names as LOC-OBJ and REF-OBJ would have been more explicit, but XARG and YARG are inherited from the top-level nodes of the hierarchy. In fact *Spatial Relations* are subsumed by the *Relation* concept.

Given the distinctions of Vandeloise (1994), our approach can be described as geometrical-topological, multiply leveled. It can be justified by the need of considering the complex interactions between our linguistic ability and other cognitive domains, in particular the vision system (Marconi, 1997). Our model partially follows Herskovits' (and Lakoff's) ideas concerning the existence of a multiplicity of levels in lexical entries of prepositions,[5] but, differently from Herskovits (1986), where all the classes of use are explicitly represented in the lexicon at the same level, we posit that relations obtained from the ideal meaning via sense shifts form a taxonomy, where higher nodes represent less informative (i.e., less specified) relations and lower nodes represent more informative (i.e., more specified) relations. This taxonomic organization of senses may overcome the problem of considering certain uses more representative than others, whereas representativeness is usually very hard to determine. For example in:

(2) a. il libro nella scatola
         the book in the box
     b. il cane nel prato
         the dog in the meadow

we assume that neither (2a) is more representative than (2b) nor the other way around. In fact, they are both derived from a general relation like Herskovits' (1986) *inclusion of a geometric object in a bi-, tri-dimensional object* (represented as *loc_in* in Fig. 4.2), but they are exactly at the same level in the taxonomy (see Fig. 4.2): Both of them specify one of the features left undetermined in the ideal meaning, that is, the dimensionality of the reference object.

In the case of locative expressions we tried to devise a set of relations, some of which stem from the necessity of coping with inferences, others from the necessity of coping with referential abilities, consisting in the ability of relating words to objects, events, and states of affairs in the world (see Marconi, 1997, for a discussion). Moreover, the more specific the relations resulting from the interpretation of prepositions are, the more accurate is, in principle, the understanding of the scene. A similar approach can be found in Olivier and Tsujii's (1994) analysis of projective prepositions, where the interpretation of the locative expression is eventually accomplished by building a depictive representation of the expression. It is clear that, to build such an "image" of the scene described in the expression, referential properties of prepositional relations must be

---

[5]This position is also compatible with the Cognitive Grammar's claim that polysemy is the norm and that networks of senses are necessary to account for lexical meanings (see Langacker, 1988).

accounted for. Anyway, it is an open question which spatial relations are actually computed by our perceptive system, how they should be defined, and which is the contribution of constraints coming from object knowledge and object location (Hayward & Tarr, 1995; Landau & Jackendoff, 1993; McNamara, 1986).

4. SEMANTIC REPRESENTATION OF LOCATIVE EXPRESSIONS

The meaning of a preposition is a binary relation that takes as arguments the interpretation of the head of the NP modified by the PP and the interpretation of the object NP. For the sake of simplicity, in the following we consider only expressions of the form NP + PP[Prep + NP], even if the model is also implemented for patterns of the type NP + VP[V + PP[Prep + NP]]. Let us consider the expression:

(3) il libro nella scatola
     the book in the box

A box is a physical object and so it identifies a portion of space: a *bounded volume*. Notice that a sentence like *the box in the book* would be rejected, even though a book can also identify a bounded volume. The difference between the volume identified by a box and the volume identified by a book is that the first is a "container" (a book can be a container only in very particular conditions). This fact can also be regarded as an indication of the necessity of considering functional contrasts such as container/contained (cf. Vandeloise, 1994). One sense shift of the preposition *in* with respect to its ideal meaning *loc_in* is *loc_in_bounded_volume* (see Fig. 4.2), with a physical object as first argument and a bounded volume as second argument. The meaning of the locative expression can be described as:

(4) $\exists x,y,z$ (book(x) & bounded_volume(y) & box(z) & spatially_identifies (z,y) & loc_in_bounded_volume (x, y))

The locative expression contains an implicit relation (i.e., without reference to a linguistic token) *spatially_identifies*, between the box and the portion of space it identifies, and an explicit relation *loc_in_bounded volume*, between the book and that portion of space.

Figure 4.3 describes (4) in a net formalism. The portion of net above the horizontal line is a fragment of the terminological knowledge; the portion below the horizontal line contains the nodes instantiated during the interpretation of the linguistic expression. The instantiated subnet in

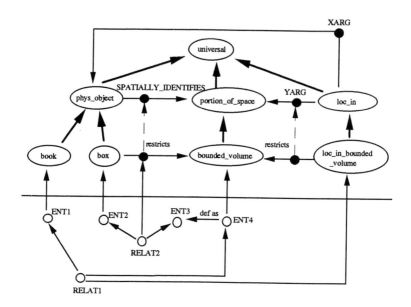

FIG. 4.3. Semantic representation of the locative expression *il libro nella scatola*.

the figure can be read as "There is a relation (RELAT$_1$) of type *loc_in_ bounded_volume* between an entity (ENT$_1$) of type *book* and another entity (ENT$_4$), which is a *bounded_volume*, defined as (*def-as*) the second argument (ENT$_3$) of a relation (RELAT$_2$) of type SPATIALLY_IDENTIFIES, whose first argument is an entity (ENT$_2$) of type *box*. This latter relation (RELAT$_2$) is the implicit relation identified during the interpretation. The arguments of RELAT$_1$ are the located object ENT$_1$ and the portion of space ENT$_4$ defined as (*def-as*) the portion of space spatially identified (RELAT$_2$) by the reference object ENT$_2$.

5. INTERPRETATION OF LOCATIVE EXPRESSIONS

The interpretation process is incremental, in the sense that the semantic representation is built piecemeal as soon as the linguistic expressions are analyzed syntactically. It takes into account the specific linguistic context in which the preposition occurs and specializes the ideal meaning(s) as far as possible.

Lexical ambiguity, concerning relations that belong to different taxonomies (homonymy), is dealt with through the introduction of a structured representation, called *prepositional ambiguity space* (PAS). A PAS initially contains instantiations of all ideal meanings expressed by a preposition, namely the roots of the taxonomies accessible from the lexical entry (see

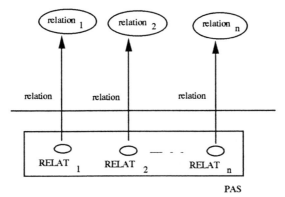

PAS

FIG. 4.4. Representation of lexical ambiguity via a PAS (prepositional ambiguity space).

Fig. 4.4). Subsumption relation permits the access to more specific (i.e., sense-shifted) relations. Disambiguation occurs when the meaning of the preposition is composed with the meanings of the NPs.

Specialization (sense shifting) of a relation R concerns relations in a single taxonomy (polysemy) and it is the process of retrieving a more specific relation R_i subsumed by R, given its actual arguments. In the terminological net there must be a "path" between the concept acting as value restriction of the role ARG of R (called ValueRestrictionConcept) and the concept associated with the filler of the argument ARG according to the syntactic structure (FillerConcept). Two cases are possible:

1. FillerConcept is directly subsumed by ValueRestrictionConcept.
2. It is possible to find another concept (called RelatedConcept), subsumed by ValueRestrictionConcept, that is connected via some role to FillerConcept. In this case we assume that there exists, for each spatial relation, a *characteristic* role to be traversed in order to identify the type of argument of the relation.

For example, referring to Fig. 4.2, the concept *book* can be the Filler-Concept of XARG of *loc_in* because it is subsumed by *physical-object*; the concept *box* can be the FillerConcept of YARG of *loc_in* because it is connected via the characteristic role of *loc_in* (SPATIALLY_IDENTIFIES) to a RelatedConcept (*bounded_volume*) subsumed by *portion_of_space*. In general, a locative expression requires the first type of check for XARG and the second for YARG.[6]

---

[6]There are some exceptions, like the expression *la zona del porto* ("the area of the harbor"), where the portion of space is explicitly mentioned. These cases are treated by letting the type of check depend on the type of argument (*box* vs. *area*).

The interpretation of a locative expression of type $NP_1 - Prep - NP_2$ consists of five steps: (a) interpretation of NP_1 and construction of the corresponding semantic representation S_1; (b) access to the meaning of the preposition and construction of a corresponding PAS, which contains all the relations retrieved from the lexical entry; (c) interpretation of NP_2 and construction of the corresponding semantic representation S_2; (d) interpretation of the pattern $Prep + NP_2$, which tests the compatibility of each relation in the PAS with S_2; incompatible relations are deleted from the PAS; compatible relations are specialized as far as possible; and (e) composition of S_1 with the representation produced in the step d, while determining possible further disambiguation.

The algorithm is not strictly incremental. Indeed, the interpretation of the PP, that is, $Prep + NP_2$, precedes the interpretation of the attachment, that is, $NP_1 + Prep$. This strategy has many computational advantages in the resolution of lexical ambiguity. Empirical data, on both the configuration $NP_1 + Prep + NP_2$ and $Verb + Prep + NP_2$, show that the pattern $Prep + NP_2$ constrains the meaning more than the patterns $NP_1 + Prep$ and $Verb + Prep$. The segmentation of attachments we propose ($Prep + NP_2$ followed by $NP_1 + Prep$ or $Verb + Prep$) provides many computational gains over the fair constraint-based model (Eberhard et al., 1995) and the strict incremental segmentation (Steedman, 1989). In strategies ($NP_1 + Prep$ or $Verb + Prep$ followed by $Prep + NP_2$) there are several cases of intractability, especially due to the high vagueness and ambiguity of prepositional and, in some cases, verbal meanings. Here follow some examples:

(5) andare (*go*) per . . .
$$\begin{cases} \text{. . . funghi } \textit{for mushrooms} \text{ (}\textit{go looking for}\text{)} \\ \text{. . . un amico } \textit{for a friend} \text{ (}\textit{beneficiary}\text{)} \\ \text{. . . i prati } \textit{across the fields} \\ \\ \text{. . . tre ore } \textit{for three hours} \\ \text{. . . una strada stretta } \textit{through a narrow street} \\ \text{. . . la fretta } \textit{for being in a hurry} \end{cases}$$

(6)
$$\left.\begin{array}{l} \text{andare } \textit{go across} \\ \text{fondi } \textit{funds} \\ \text{strada } \textit{road} \end{array}\right\} \text{per Torino } \textit{for/to Torino}$$

The following algorithm performs compatibility checks and relation specialization invoked in steps d and e.[7]

---

[7] In Herskovits (1986) the specialization process does not occur, because the senses of a preposition are not organized in a taxonomy and are collectively enumerated in the lexical entry (as classes of use).

Check&Specify (*Relation, Role, Meaning*)
let CharacteristicRole = the characteristic role associated with Relation for
   that Meaning
if CharacteristicRole = nil
then Node =
   if ValueRestriction (Role, Relation) subsumes Meaning
   then Meaning
   else nil
else Node =
   if ValueRestriction (Role, Relation) subsumes ValueRestriction (Charac-
      teristicRole, Meaning)
   then ValueRestriction (CharacteristicRole, Meaning)
   else nil
if Node = nil
then Remove (Relation, PAS)
else
   let Newnodes = list of nodes on the subsumption path from Node to Value-
      Restriction Role, Relation
   MostSpecificRelation = the most specific relation subsumed by Relation
   such that there is a node Node in Newnodes such that Node = Value-
   Restriction (Role, MostSpecificRelation)
   Replace (Relation, MostSpecificRelation, PAS)

The input parameters are: the relation to check and specialize (Relation),
the current role (Role = XARG or YARG), and the FillerConcept (Meaning);
the PAS contains the relations obtained from the lexical entry of the
preposition. CharacteristicRole is the characteristic role of Relation (mentioned previously). In locative expressions CharacteristicRole is usually a
restriction of SPATIALLY_IDENTIFIES or GEOMETRIC_DESCRIPTION.
   If no CharacteristicRole exists, the procedure computes the most specific concept between ValueRestrictionConcept (obtained in *Check&Specify*
from ValueRestriction(Role, Relation)) and FillerConcept (the input parameter Meaning); otherwise, the procedure computes the most specific
concept between ValueRestrictionConcept and RelatedConcept (obtained
in *Check&Specify* from ValueRestriction(CharacteristicRole, Meaning)). If
in both cases the subsumption does not hold, Relation is rejected and
removed from the PAS; otherwise, the specialization process starts along
the path between Node and ValueRestrictionConcept and, traversing the
path from the most specific concept, it looks for a relation MostSpecific-
Relation subsumed by the initial Relation that has a role whose value
restriction is the concept currently visited. MostSpecificRelation (if it exists) is a specialization of the initial Relation and replaces it in the PAS.
   Referring to Fig. 4.2, we may follow the execution of the procedure
when called on Relation=*loc_in*, Role=YARG, Meaning=*box*. The characteristic role of *loc_in*, SPATIALLY_IDENTIFIES, is saved in Charac-

teristicRole. The second branch of the first conditional statement is followed, so that it is verified whether *box* is acceptable as YARG of *loc_in*. Because ValueRestriction (YARG,*loc_in*) = *portion_of_space* and ValueRestriction (SPATIALLY_IDENTIFIES, *box*) = *bounded_volume*, it is checked whether *portion_of_space* subsumes *bounded_volume*. Because the test is successful, Node is set to *bounded_volume* and the second branch of the next conditional statement is followed, to specialize the relation in input (*loc_in*).[8] The path from *bounded_volume* to *portion_of_space* is saved in NewNodes = (*bounded_volume, volume, portion_of_space*). Starting from the most specific node in NewNodes, the path is searched for a node that restricts the value of the role YARG of a relation subsumed by *loc_in*. In the example net the relation *loc_in_bounded_volume* is found immediately. Such a relation replaces the initial *loc_in*, because it is its specialization required by the context.

6. AN EXAMPLE

Let us follow the interpretation of the following expression:

(7) il libro nella scatola
     the book in the box

1. The phrase "il libro" (*the book*) is interpreted. An instantiation of the concept *book* is introduced in the representation (ENT_1 in Fig. 4.5).

2. The lexical entry of "in" identifies four basic meanings: The interpreter creates a new PAS (PAS_1) including four relational nodes ($RELAT_1$, $RELAT_2$, $RELAT_3$, $RELAT_4$).[9]

3. The phrase "la scatola" (*the box*) is interpreted (ENT_2 in Fig. 4.6).

4. Then, "nella scatola" (*in the box*) is interpreted. The procedure *Check&Specify* is entered four times with arguments $RELAT_i$ (Relation, i.e., a prepositional sense in PAS_1), YARG (Role), and *box* (Meaning, i.e., the sense of "scatola"). The characteristic role of the four relations in PAS_1 is SPATIALLY_IDENTIFIES, which, for the concept *box*, is restricted to *bounded_volume* (see Fig. 4.2). As the result of the execution, the relations

---

[8] If the concept under analysis is not subsumed by *portion_of_space*, Node is set to *nil*, and the first branch of the next conditional statement is followed, rejecting the relation; in fact, its selectional restriction is not fulfilled.

[9] Now it should be possible to try to link the relations in the PAS with ENT_1, but in most cases no substantial changes would occur: So, we wait to specialize (and possibly reject) some relations by composing the object NP (see the earlier discussion of examples (5) and (6)).

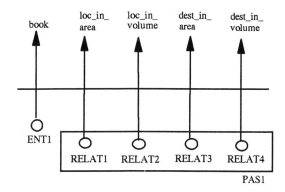

FIG. 4.5. Analysis of *il libro nella scatola* after steps 1 and 2 of the algorithm *Check&Specify*.

RELAT$_1$ (*loc_in_area*) and RELAT$_3$ (*dest_in_area*) are excluded, because *bounded_volume* is not subsumed by *area*.

The other two relations, RELAT$_2$ and RELAT$_4$, are specialized and produce *loc_in_bounded_volume* and *dest_in_bounded_volume* respectively (see Fig. 4.6). Two new nodes, RELAT$_5$ and ENT$_3$, are introduced to represent the implicit relation *spatially_identifies* and the implicit concept *bounded_volume* and the PAS is connected with ENT$_3$ with the arc YARG.

5. Finally "libro+in" is interpreted. The procedure *Check&Specify* is invoked on the meaning *book* as the filler of the role XARG for the relations still in the PAS. The relation RELAT$_4$ (*dest_in_bounded_volume*) is deleted, because its restriction on XARG requires a *motion_event*. The relation RELAT$_2$ (*loc_in_bounded_volume*) remains as the only possible interpretation (Fig. 4.7).

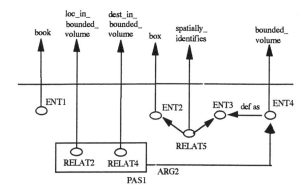

FIG. 4.6. Analysis of *il libro nella scatola* after steps 3 and 4 of the algorithm *Check&Specify*.

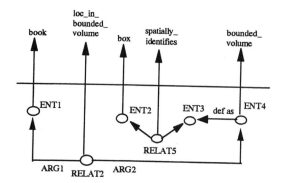

FIG. 4.7. Analysis of *il libro nella scatola* after step 5 of the algorithm *Check&Specify*.

Of course, in truly ambiguous examples, like

(8) Cristina corre nel prato
     Cristina is running into/within the field

at the end of the interpretation process the PAS contains more than one relation.

7. CONCLUSIONS

The chapter has presented a computational model for the interpretation of static locative expressions. In particular we have focused on the definition and representation of the meaning of prepositions and on how prepositions contribute to the meaning of the whole expression. The meaning of prepositions is a binary relation, but many relations can be associated with a preposition (homonymy) and most of them can be further specialized (polysemy). We have developed a geometrical/topological, multiple-level theory of prepositional meaning that tries to account for both the inferential and referential lexical competence. The computational model accomplishes both the tasks of disambiguating and specializing the most appropriate relation, given the context of use.

The model, which is implemented in Common Lisp, has been extensively tested on the static uses of the Italian prepositions *a* (*at, in, . . .*), *in* (*in*), and *su* (*on, over, above*), and has been partially tested on the dynamic uses of the Italian prepositions *a* (*to*), *in* (*to, into*), *per* (*towards, across, through*), and *da* (*from, to*).

ACKNOWLEDGMENTS

We wish to thank Diego Marconi, Pier Marco Bertinetto, and Luca Dini, who read previous versions of this chapter.

REFERENCES

Bennett, D. C. (1975). *Spatial and temporal uses of English prepositions: An essay in stratificational semantics*. London: Longman.

Boggess, L. C. (1979). *Computational interpretation of English spatial prepositions*. Unpublished doctoral dissertation, University of Illinois, Urbana.

Brachman, R. J., & Schmolze, J. G. (1985). An overview of the KL-ONE knowledge representation system. *Cognitive Science, 9*, 171–216.

Cooper, G. S. (1968). *A semantic analysis of English locative prepositions* (Bolt, Beranek, & Newman, Report No. 1587). Springfield, VA: Clearing House for Federal, Scientific, and Technical Information.

Eberhard, K., Spivey-Knowlton, M., Sedivy, J., & Tanenhaus, N. (1995). Eye movement as a window into real-time spoken language comprehension in natural contexts. *Journal of Psycholinguistic Research, 24*, 409–436.

Hawkins, B. (1988). The category "medium." In B. Rudzka-Ostyn (Ed.), *Topics in cognitive linguistics* (pp. 231–270). Amsterdam and Philadelphia: John Benjamins.

Hayward, W. G., & Tarr, M. J. (1995). Spatial language and spatial representation. *Cognition, 55*, 39–84.

Herskovits, A. (1986). *Language and spatial cognition*. Cambridge, England: Cambridge University Press.

Herweg, M. (1989). Ansätze zu einer semantischen Beschreibung topologischer Präpositionen [A preliminary analysis of the semantics of topological prepositions]. In C. Habel, M. Herweg, & G. Rehkämper (Eds.), *Raumkonzepte in Verstehensprozessen* (pp. 99–127). Tubingen, Germany: Niemeyer.

Jackendoff, R. (1983). *Semantics and cognition*. Cambridge, MA: MIT Press.

Jackendoff, R. (1990). *Semantic structures*. Cambridge, MA: MIT Press.

Japkowicz, N., & Wiebe, J. M. (1991, June). A system for translating locative prepositions from English into French. In *Proceedings of the 23rd Annual Meeting of the Association for Computational Linguistics, Berkeley, CA* (pp. 153–160).

Johnson-Laird, P. N. (1983). *Mental models*. Cambridge, England: Cambridge University Press.

Kalita, J., & Badler, B. (1991, July). Interpreting prepositions physically. In *Proceedings of 9th National Conference on Artificial Intelligence, Anaheim, CA*. Cambridge, MA: AAAI Press/MIT Press.

Lakoff, G. (1987). *Women, fire and dangerous things. What categories reveal about the mind*. Chicago: Chicago University Press.

Landau, B., & Jackendoff, R. (1993). "What" and "Where" in spatial language and spatial cognition. *Behavioral and Brain Sciences, 16*, 121–141.

Lang, E. (1993). The meaning of German projective prepositions: A two-level approach. In C. Zelinsky-Wibbelt (Ed.), *The semantics of prepositions: From mental models to natural language processing* (pp. –). Berlin: Mouton de Guyter.

Langacker, R. W. (1988). A view of linguistic semantics. In B. Rudzka-Ostyn (Ed.), *Topics in cognitive linguistics* (pp. 49–90). Amsterdam and Philadelphia: John Benjamins.

Langacker, R. W. (1991). *Concept, image, symbol*. New York: Mouton de Gruyter.

Marconi, D. (1997). *Lexical competence*. Cambridge, MA: MIT Press.

McNamara, T. P. (1986). Mental representations of spatial relations. *Cognitive Psychology, 18*, 87–121.

Miller, G. A., & Johnson-Laird, P. N. (1976). *Language and perception*. Cambridge, MA: Harvard University Press.

Olivier, P. L., & Tsujii, J. (1994, June). A computational view of the cognitive semantics of spatial prepositions. In *Proceedings of the 32nd Annual Meeting of the Association for Computational Linguistics, Las Cruces, NM* (pp. 303–309).

Pustejovsky, J. (1991). The generative lexicon. *Computational Linguistics, 17*, 409–441.

Quine, W. V. O. (1951). Two dogmas of empiricism. *Philosophical Review, 60*, 20–43. (Later in *From a logical point of view*. Cambridge, MA: Harvard University Press, 1953)

Steedman, M. J. (1989). Grammar, interpretation and processing from the lexicon. In W. Marslen-Wilson (Ed.), *Lexical representation and process* (pp. 463–504). Cambridge, MA: MIT Press.

Talmy, L. (1985). Lexicalization patterns: Semantic structures in lexical forms. In T. Shopen (Ed.), *Language typology and syntactic description: Vol. 3. Grammatical categories and the lexicon* (pp. 57–149). Cambridge, MA: Cambridge University Press.

Vandeloise, C. (1994). Methodology and analyses of the preposition "in." *Cognitive Linguistics, 5*, 157–184.

Waltz, D. L. (1980). Generating and understanding scene descriptions. In A. Joshi et al. (Eds.), *Elements of discourse understanding* (pp. 266–282). Cambridge, England: Cambridge University Press.

Zwicky, A. M., & Saddock, J. M. (1975). Ambiguity tests and how to fail them. In J. P. Kimball (Ed.), *Syntax and semantics 4* (pp. 1–36). New York: Academic Press.

5

Generating Dynamic Scene Descriptions

Hilary Buxton
Richard J. Howarth
University of Sussex

This chapter addresses the problem of extracting descriptions of object behavior from image sequences. Vision systems are now capable of delivering trajectory-based descriptions of moving objects in a scene but little work has been done on the spatio-temporal reasoning needed for the computation of behavioral descriptions. This level of understanding allows us to compute meaningful descriptions of what is happening in a scene. We have developed analogical, cellular representations of space and time together with deictic descriptions of the behavior of moving objects to support the generation of this kind of dynamic scene description. In addition, we propose that an active, purposive framework is required for advanced vision as we need to be selective and deliver a situated (here and now) analysis of behavior in the scene. The problem of dynamic surveillance of traffic scenes is used for illustration.

1. INTRODUCTION

In advanced vision systems we often need knowledge-based techniques that, although not completely general, are effective in a particular domain. For example, early work by Brooks (1981) used symbolic reasoning to analyze static airport scenes in a cycle of prediction, description, and interpretation. Such explicit reasoning about constraints is unable to deliver the fast performance required in dynamic scene interpretations. Recent attempts to improve the reliability and computational tractability of such systems have been based on applying Bayesian belief networks and decision theories (Binford, Levitt, & Mann, 1989; Levitt, Binford, & Ettinger, 1990). However, these studies do not address the specific computational difficulties involved in the interpretation of image sequences. Recent work using model-based approaches in image sequence analysis (Koller, Daniilidis, Thorhallson, & Nagel, 1992; Worrall, Marslin, Sullivan, & Baker, 1991) do effectively address the issue of dynamic interpretation. However, they do not compute behavioral descriptions or use task-dependent processing. Here we describe dynamic knowledge-based processing in an active, selective framework to address both these issues.

We can adapt processing to the task at hand by using database query techniques; for example, Buxton and Walker (1988) proposed a scheme for incorporating explicit spatio-temporal knowledge into a query-based vision system for understanding biological image sequences. However, this kind of system performs offline processing and uses a fixed set of parameters for the object(s) of interest. Here, however, we want to interpret the spatio-temporal interactions between observed objects online while using the scene as its own best memory. To build such vision systems we need to address the question of how knowledge can be mapped onto computation to dynamically deliver consistent interpretations. This involves a more fundamental analysis of the spatio-temporal regularities in the image data so that we can exploit them as constraints in the processing scheme. We recently developed a Bayesian network approach (Gong & Buxton, 1993) as it can support this kind of effective knowledge representation. It also provides the means for solving the information integration problem, which is central to building robust systems capable of working on real image sequence data.

The issues involved in using such active and purposeful strategies in artificial systems are well discussed elsewhere. Our approach owes much to Ballard (1991), who proposed an "animate vision" approach, for two reasons: First, vision is better understood in the context of the visual behaviors engaging the system without requiring detailed internal representations of the scene; and second, it is important to have a system framework that integrates visual processing within the task context. Many researchers have now integrated such purposive vision systems (Aloimonos, Weiss, & Bandopadhay, 1988; Rimey & Brown, 1992; Tsotsos, 1992). Here we are addressing the core issues of "what" should be looked for, "where" and "when" we need to attend in order to efficiently compute behavioral descriptions.

2. DYNAMIC SCENE DESCRIPTIONS

To compute dynamic scene descriptions, a simple notion of time and events is required in an online vision system. Some researchers, for example, Brooks (1991), have even suggested that we can dispense with all internal representations and allow the world to be its only model using reactive control for intelligent activity. However, we propose to go only part way toward this approach with a limited situated analysis in which we represent the properties that are relevant for our visual tasks. These properties will enable us to identify when a change occurs in a meaningful context. For example, in surveillance, this context is the representation of our ground-plane (scene) knowledge and other current purposively moving objects. For instance, Fig. 5.1 illustrates three images in which a lorry and a few cars interact on a roundabout.

Nagel (1988) reviewed the few projects that have tackled the problem of delivering conceptual descriptions in the road traffic domain. These include NAOS (Neumann, 1989) and CITYTOUR (Retz-Schmidt, 1988), which allow question answering as an offline query process. These descriptions require composition by grammar and much of this work has used the logico-linguistic approach of Miller and Johnson-Laird (1976) and analysis of spatial prepositions by Herskovits (1986). Nagel also considered the problem of online generation of such descriptions. His approach goes some way toward the goal of specifying conceptual descriptions in terms of motion verbs that could be effectively computed. However, we think a more situated approach has advantages as it uses a perceiver-centered or "deictic" frame of reference. For example, spatial deixis uses words like *here* and *there* and temporal deixis words like

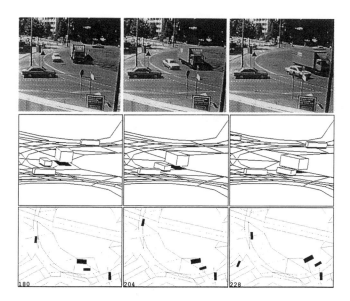

FIG. 5.1. The image plane, three-dimensional space with dynamic objects, and the ground-plane representation of a traffic roundabout.

yesterday and *now*. Deictic reference depends on knowledge of the context but can decompose and simplify the reasoning that needs to be done compared to the global "state-based" approach of traditional artificial intelligence (AI) and seems deeply embedded in our communication and spatial reasoning (Buhler, 1982; Retz-Schmidt, 1988).

We require a more local frame in surveillance than that required to support full cognitive planning or coordinating actions as we are only observing the activity of moving objects. Formal logic approaches using well-defined languages with clear meaning for time, events, and causality; for example, Allen (1984) and Shoham (1988) are useful for validating and prototyping new approaches to behavioral analysis but do not seem suitable for online vision systems. To effectively mix both qualitative and quantitative descriptions of space and time for a wide set of visual tasks, we used Fleck's (1986, 1988) cellular models, which discretize real time and space. For example, Fig. 5.2 illustrates the cellwise spatial decomposition of the roundabout ground-plane.

In cellwise time, as developed by Fleck (1986, 1988), each cell is a state and we can classify change as either: "state-changes," which involve sharp

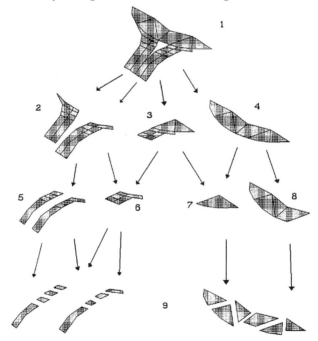

FIG. 5.2.   Hierarchical decomposition into regions and cells representing a part of the roundabout ground-plane. Reprinted from Fig. 5 of Haworth and Buxton (1992), Copyright © 1992, with kind permission of Elsevier Science—NL, Sara Burgerhartstrat 25, 1055 KV Amsterdam, The Netherlands.

change; "activities," where continuous change occurs; or "accomplish-ments," which are composites of activity and state-change. "Episodes" are seen as composed of a starting state-change, an activity, and an end state-change in this framework. This kind of representation can be called "analogical" and has been further developed for the representation of events and behavior under task-based control for surveillance (Howarth & Buxton, 1992, 1993).

3. BAYESIAN TECHNIQUES

To implement our active behavioral analysis, we use the Bayesian belief revision approach that is conceptually attractive and computationally feasible for vision (Levitt et al., 1990; Nicholson & Brady, 1992; Rimey & Brown, 1992). In an online system simple correlations in the spatio-tem-poral data can be exploited to efficiently infer quite complex behavior. For example, Fig. 5.3 illustrates the traffic roundabout scene and correlated vehicle trajectories.

Modeling and updating the dependent relationships and their prob-ability distributions in belief nets is relatively easy both offline or online. We have demonstrated this for both motion segmentation and tracking (Gong & Buxton, 1993) and in the evaluation of visual behavior (Howarth & Buxton, 1993). Conceptually, we are addressing the issue of modeling an information retrieval process that purposively collects evidence in the

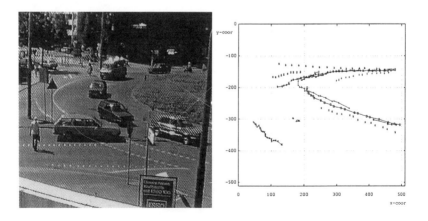

FIG. 5.3. Left: a traffic roundabout scenario and its traffic flow. Right: correlated spatio-temporal constraints on the movements of individual objects are imposed implicitly by this scene layout. From Gong and Buxton (1993). Reprinted with permission of BMVA Press.

image to support interpretations of dynamic behaviors in the scene. Ambiguities in individual levels of computation mean that context-dependent information integration is required to obtain more coherent interpretations of the visual evidence. Recent developments in probabilistic relaxation, belief, and decision theory have provided us with a sound computational base (Pearl, 1988), which we can extend for the problem at hand.

Bayesian belief networks are directed acyclic graphs (DAGs) in which each node represents an uncertain quantity using variables. For example, Fig. 5.4 illustrates the relationship between scene layout and flow vector measures from the image sequence.

The arcs connecting the nodes signify the direct causal influences between the linked variables with the strengths of these influences quantified by associated conditional probabilities. We use only singly connected trees for modeling as these are fast to update. A selection of these multivalued variables will be the direct causes (parents) at a particular node. The strengths of these direct influences are quantified by assigning a link matrix for every combination of values of the parent set. The conjunction of all the local link matrices of variables in the network specifies a complete and consistent global model, which is given by the overall joint distribution function over the variable values. The behavior of a visual process is partially defined by its processing parameters, which are updated in the network so dynamic evaluation will be consistent with the visual task.

In a belief network, we can quantify the degree of coherence between the expectations and the evidence by a measure of local belief and define

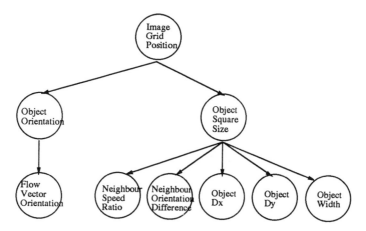

FIG. 5.4. A belief network that captures the dependent relationships between the scene layout and relevant measures in motion segmentation and tracking. From Gong and Buxton (1993). Reprinted with permission of BMVA Press.

belief commitments as the tentative acceptance of a subset of hypotheses that together constitute a most satisfactory explanation of the evidence at hand. Bayesian belief revision updates belief commitments by distributed local message passing operations. Instead of associating a belief measure with each individual hypothesis, belief revision identifies a composite set of hypotheses that best explains the evidence. We call such a set the most probable explanation (MPE). The conceptual basis of the propagation mechanism that updates the network is quite simple. For each hypothetical value of a single variable, there exists a best extension of the complementary variables. The problem of finding the best extension can be decomposed into finding the best complementary extension to each of the neighboring variables according to their conditional dependencies. This information can then be used to decide the best value at the node. The decomposition allows the processing to be applied recursively until it reaches the network boundary where evidence variables have predetermined values.

4. BEHAVIORAL EVALUATION

The Bayesian approach can be used to overcome the problem of uncertainty and incompleteness in the evaluation of behavior by bringing task-based and scene-based knowledge into the interpretation process. Dynamic scene interpretation in surveillance requires both modeling what one is looking for (top-down expectations) and interpreting evidence of what could be appearing (bottom-up inference). The prior probabilities can be used to initialize the network and then the evidence is dynamically interpreted under the current expectations using both parent to child (top-down) and child to parent (bottom-up) updating of values in the network. We have effectively used such Bayesian networks together with a deictic representation both to create a dynamic structure to reflect the spatial organization of the data and to measure task relatedness. Howarth and Buxton (1993) integrated the behavioral evaluation and interpretation by giving a combined attentional focus for the road traffic exemplar where the behaviors of interest were "overtaking," "following," "queuing," and "unknown." For example, a simple proximity cue invokes the behavioral analysis of overtaking. A task-based Bayesian network (adapted from Rimey & Brown, 1992) is used in modeling spatial and temporal relationships in order to direct the evidence collection in the image sequence.

We use a separation of preattentive (peripheral) and attentional processing in our behavioral evaluation system. The simple peripheral operators such as velocity and proximity on the ground-plane act as cues for the more complex attentional operations such as path prediction and computing deictic spatial relationships that are used in the full evaluation. An attentional mechanism guides the application of appropriate complex evaluation in a particular dynamic context and makes the reasoning relevant to the current task. This attentional mechanism uses an agent-based formalism implemented by Bayesian network updating. The objects in our scenes have an "intrinsic front," which defines each object's frame of reference in the deictic representation. We have developed "typical-object-models" for the interpretation of behavior using time-ordered combinations of deictic relationships between objects of interest in particular ground-plane contexts. The deictic relationships may be simply ("behind" "previous"), ("beside" "now"), and ("infront" "next"). The simple peripheral operators are applied to all our segmented, tracked objects and the typical-object-model determines the specific attentional operations to be performed. It also determines which values should be saved and the set of operations to be performed on the next clock tick. The results are fed back to the appropriate agent to give task-related features for future selection. The approach here is related to Agre and Chapman (1987) but extended to deal with several local deictic viewpoints. In this way an agent need not describe every object in the domain but only those relevant to its particular task.

To combine the information that develops over time we use a dynamic form of Bayesian network (DBN), which captures the changing relationships between scene objects. The DBN is composed of temporally separated subgraphs that are interconnected using reconfigurable links. The node building and node linking updates the structure for the current time so that a new tree is obtained that inherits values for beliefs at nodes referenced in the "official observer" view. These markers, maintained centrally by the official observer, solve the problem of consistently associating entities that are maintained only locally by the distributed agents. A matrix of conditional probabilities captures the interest of the proximity relationships according to whether the objects are: "not-near," "nearby," "close," "very-close," or "touching." If, at the current time point, A and B remain near each other, we form the temporal links and extract a new tree structure rather than a multiply connected network. In the example here, each tree holds values from only the current and immediately previous time points and a tree is formed for each relationship so it is trivial to prove that no loops are introduced, which is important for bounding the computation. Once all propagations are complete, the most interesting relationship can be selected.

The structural changes in the DBN reflect the simple, monitored relationships between all objects of interest. For those objects involved in the most interesting relationships, this DBN is augmented by a "tasknet," which is a structurally fixed Bayesian network that builds a coherent interpretation of the temporally evolving behavior. The input nodes here represent key features relevant to the task it has been constructed to identify and the output root node represents the overall belief in the behavioral task based on evidence collected so far. Allocating an attentional process explicitly in this way ensures that a tasknet is running on the selected objects and can terminate when an uninteresting situation is recognized. The attentional system has three properties: "focus-of-attention," which ensures only the target hypothesis and associated functions are updated; "terminate-attention," which can stop all activities associated with a target hypothesis that has been confirmed or denied to an acceptable degree of confidence; and "selective-attention," which allows dynamic selection of the most interesting hypothesis to watch.

5. CONCLUSION AND FUTURE WORK

This work has discussed active behavioral analysis of dynamic scenes (see also Howarth, 1994). The aim is to provide timely behavioral descriptions by using appropriate representations with associated reasoning on the image sequences for surveillance tasks. Similar work by Rimey and Brown (1994) on the control of selective perception when using an active camera system has also used Bayesian networks together with decision theory for static problems such as identifying what form of place setting has been laid out for dinner. In future work we hope to use an active camera platform (such as that described by Murray, 1993), and use the official observer's analogical ground-plane representation of the scene as a short-term memory of where objects are and how they are moving. The camera system will then be able to physically look away from an object and back again when it is again of interest. The local deictic representations of the objects and their interactions should also tie into the active camera system's current view direction to simplify the mapping of the observer- and object-centered frames of reference. This kind of integration of behaviorally driven attention will allow us, for the first time, to develop intelligent sensing strategies in space and time for dynamic scenes.

Another aspect of the Bayesian network framework, which is crucial for the success of advanced vision systems, is the ability to encode knowledge. We discussed modeling dynamic dependencies among the visual parameters in space and time, but there is also a need to model the prior probabilities of classes of interpretation to stabilize the initial estimation. It is possible to provide these by analysis of the problem where there are obvious scene-based constraints such as traffic flow direction in certain lanes of a roundabout. It may also be possible, in future work, to learn these constraints and dependencies using appropriate techniques (Spiegelhalter & Cowell, 1992; Whitehead & Ballard, 1991). This can be a time-consuming process but is typically computed offline with only limited adaptive refinement online. If we can reliably learn the salient behavioral properties from the scene and objects by observation, a new era of advanced autonomous vision systems would become feasible.

ACKNOWLEDGMENTS

We are grateful to Shaogang Gong for the illustrations used in this chapter for Figs. 5.3 and 5.4. The work reported in this chapter was funded by EPSEC grant GR/K08772.

REFERENCES

Agre, P. E., & Chapman, D. (1987). Pengi: An implementation of a theory of activity. In *Proceedings of National Conference on Artificial Intelligence* (pp. 268–272). Menlo Park, CA: AAAI Press.

Allen, J. F. (1984). Toward a general theory of action and time. *Artificial Intelligence, 23,* 123–154.

Aloimonos, Y., Weiss, I., & Bandopadhay, A. (1988). Active vision. *International Journal of Computer Vision, 1,* 35–54.

Ballard, D. H. (1991). Animate vision. *Artificial Intelligence, 48,* 57–86.

Binford, T. O., Levitt, T. S., & Mann, W. B. (1989). Bayesian inference in model-based machine vision. In L. N. Kanal, T. S. Levitt, & J. F. Lemmer (Eds.), *Uncertainty in artificial intelligence 3* (pp. 207–221). New York: North-Holland.

Brooks, R. A. (1981). Symbolic reasoning among 3D models and 2D images. *Artificial Intelligence, 17,* 285–348.

Brooks, R. A. (1991). Intelligence without reason. In *Proceedings of International Joint Conference on Artificial Intelligence* (pp. 569–595). New York: Morgan Kaufmann.

Buhler, K. (1982). The deictic field of language and deictic words. In R. J. Jaruella & W. Klein (Eds.), *Speech, place and action: Studies in deixis and related topics* (pp. 9–30). New York: Wiley.

Buxton, H., & Walker, N. (1988). Query-based visual analysis: Spatio-temporal reasoning in computer vision. *Image and Vision Computing, 6,* 247–254.

Fleck, M. (1986). Representing space for practical reasoning. *Image and Vision Computing, 6,* 75–86.

Fleck, M. (1988). *Boundaries and topological algorithms.* Unpublished doctoral dissertation, MIT, Cambridge, MA.

Gong, S. G., & Buxton, H. (1993). Bayesian nets for mapping contextual knowledge to computational constraints in motion segmentation and tracking. In J. Illingworth (Ed.), *Proceedings of British Machine Vision Conference* (pp. 229–238). Guildford, Surrey, England: BMVA Press.

Herskovits, A. (1986). *Language and spatial cognition: An interdisciplinary study of the prepositions in English.* Cambridge, England: Cambridge University Press.

Howarth, R. (1994). *Spatial representation, reasoning and control for a surveillance system.* Unpublished doctoral dissertation, University of London, London.

Howarth, R., & Buxton, H. (1992). An analogical representation of space and time. *Image and Vision Computing, 10,* 467–478.

Howarth, R., & Buxton, H. (1993). Selective attention in dynamic vision. In *Proceedings of International Joint Conference on Artificial Intelligence* (pp. 1579–1584). New York: Morgan Kaufmann.

Koller, D., Daniilidis, K., Thorhallson, T., & Nagel, H.-H. (1992). Model-based object tracking in traffic scenes. In *Proceedings of European Conference on Computer Vision* (pp. 437–452). New York: Springer-Verlag.

Levitt, T. S., Binford, T. O., & Ettinger, G. J. (1990). Utility-based control for computer vision. In M. Henrion, R. D. Shachter, L. M. Kanal, & J. F. Lemmer (Eds.), *Uncertainty in artificial intelligence* (pp. 371–388). New York: North-Holland.

Miller, G., & Johnson-Laird, P. N. (1976). *Language and perception.* Cambridge, MA: Harvard University Press.

Murray, D. W. (1993). Reactions to peripheral image motion using a head/eye platform. In *Proceedings of International Conference on Computer Vision* (pp. 403–441). Los Alamitos, CA: IEEE Press.

Nagel, H. H. (1988). From image sequences towards conceptual descriptions. *Image and Vision Computing, 6,* 59–74.

Neumann, B. (1989). Natural language description of time varying scenes. In D. L. Waltz (Ed.), *Semantic structures* (pp. 167–206). Hillsdale, NJ: Lawrence Erlbaum Associates.

Nicholson, A. E., & Brady, J. M. (1992). The data association problem when monitoring robot vehicles using dynamic belief networks. In *Proceedings of European Conference on Artificial Intelligence* (pp. 689–693). New York: Springer-Verlag.

Pearl, J. (1988). *Probabilistic reasoning in intelligent systems, networks of plausible inference.* New York: Morgan Kaufmann.

Retz-Schmidt, G. (1988). Various views on spatial prepositions. *AI Magazine, 9*(2), 95–105.

Rimey, R. D., & Brown, C. M. (1992). Where to look next using a Bayes net: Incorporating geometric relations. In *Proceedings of European Conference on Computer Vision* (pp. 542–550). New York: Springer-Verlag.

Rimey, R. D., & Brown, C. M. (1994). Control of selective perception using Bayes nets and decision theory. *International Journal of Computer Vision, 12,* 173–208.

Shoham, Y. (1988). *Reasoning about change: Time and causation from the standpoint of artificial intelligence.* Cambridge, MA: MIT Press.

Spiegelhalter, D. J., & Cowell, R. G. (1992). Learning in probabilistic expert systems. In *Bayesian Statistics 4* (pp. 47–67). New York: Oxford University Press.

Tsotsos, J. K. (1992). On the relative complexity of active vs. passive visual search. *International Journal of Computer Vision, 7,* 127–142.

Whitehead, S. D., & Ballard, D. H. (1991). Learning to perceive and act by trial and error. *Machine Learning, 7,* 45–83.

Worrall, A. D., Marslin, R. F., Sullivan, G. D., & Baker, K. D. (1991). Model-based tracking. In P. Mowforth (Ed.), *Proceedings of British Machine Vision Conference* (pp. 310–318). London: Springer-Verlag.

6

A Three-Dimensional Spatial Model for the Interpretation of Image Data

Thomas Fuhr
Gudrun Socher
Christian Scheering
Gerhard Sagerer
Universität Bielefeld

We present a means of generating and understanding relative spatial positions in a natural three-dimensional (3-D) scene, in terms of six spatial prepositions, left, right, in-front, behind, above, and below, using real stereo images. Our model has two layers. First, a symbolic spatial description of the scene independent of reference frames is computed. Then, in the second layer, the meaning of each of the six prepositions is defined with respect to the current reference frame, based on the description from the first layer. The meaning definitions of the prepositions in the given model can be used in two ways. They allow the system to judge the degree of applicability of each of the six prepositions between two 3-D objects according to a graduated scale; and given the 3-D object description of one object, the admissible two-dimensional (2-D) image region of the other object can be inferred.

1. INTRODUCTION

Recent spatial modeling and reasoning has focused on five major topics: (a) development of efficient data structures, storage, and retrieval of spatial configurations, for example, in the field of geographical information systems; (b) purely qualitative reasoning in 2-D/3-D space by means of constraint satisfaction techniques (Guesgen, 1989; Hernández, 1993) and to support qualitative simulation processes (Cui, Cohn, & Randell, 1992); (c) how people use spatial prepositions in speech and writing

(Herrmann, 1990; Retz-Schmidt, 1988); (d) visualizing qualitatively described spatial constellations (Hernández, 1993; Olivier & Tsujii, 1994;
Schirra & Stopp, 1993); (e) systems that communicate with human users
about 2-D or 3-D spatial configurations based on numerical scene information (Abella & Kender, 1993; André, Herzog, & Rist, 1988; Gapp, 1994).

The spatial modeling approach we present in this chapter has been
designed to contribute to a system of type (e), involving both speech
understanding and computer vision. Here, a system observes the scene
with a stereo camera and communicates with a human partner via speech
in order to solve a construction task. For such a scenario, the exchange of
qualitative spatial information is essential. In particular, the system must
be able to *generate* qualitative spatial information about the scene from
numerical visual input data. It must also be able to *understand* the qualitative spatial information contained in instructions from the human partner;
that is, it must be capable of relating this information to the visual numerical
data.

We look at a part of this problem. We present a 3-D spatial model
using the spatial relations left, right, in-front, behind, above, and below,
which we call *spatial prepositions*.[1] In contrast to other approaches our aim
is to both generate and understand the spatial information contained in
simple utterances like "*the wheel (is) to the left of the long bar.*" In such
expressions the location of an object (*wheel*), which we call the located
object (LO), is given relative to another object (*long bar*), called the reference object (RO), by means of a preposition (*left*). We experiment with
objects from a children's toolbox, such as bolts, connection bars, and the
like, shown in Fig. 6.1.

Other approaches to similar problems concentrate on *either* generating
spatial information from numerical data (Abella & Kender, 1993; André
et al., 1988; Gapp, 1994) *or* understanding the meaning of a spatial
description in the sense of visualizing verbally described spatial configurations (Hernández, 1993; Olivier & Tsujii, 1994; Schirra & Stopp, 1993).
Stopp, Gapp, Herzog, Laengle, and Lueth (1994) presented a system for
bidirectional natural language communication with a robot. They addressed both the generation of prepositions for LO-RO pairs as well as
the problem of localizing a LO with respect to a given RO and some
preposition. However, in this approach, object recognition in the 2-D
images cannot profit from the qualitative localization. Hypotheses about
objects are always constructed on the basis of the complete images. A
generate-and-test mechanism then selects those pairs that best fulfill the
preposition.

---

[1]We are aware of the fact that the meaning of spatial prepositions can usually not be
modeled by single relations (Herskovits, 1986). However, our prepositions are a first
approximation to the "ideal meaning."

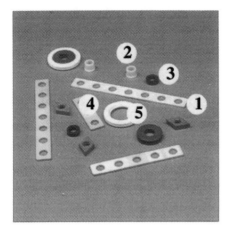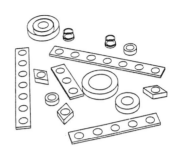

FIG. 6.1. An observed scene and its model-based 3-D reconstruction in a side view; the location of the numbered objects is analyzed in detail.

Our model is suited for bidirectional communication. Spatial prepositions are generated for pairs of 3-D objects. The object poses in 3-D are reconstructed from stereo images. What is new in this work is the ability to numerically locate the LO in the *2-D images* from the qualitative 3-D relation of the LO to the RO. Given the reconstructed pose of the RO, we compute the possible image regions where the LO can be located. Thus, the effort for recognizing the LO may be reduced because it only has to be performed in parts of the image.

Our approach differs from those previously mentioned in another manner. Prepositions are not generated *directly* from 3-D object descriptions. Rather, in the first step, an object-centered relational representation is determined. This is independent of any reference frame used in verbal communication—namely intrinsic, deictic, or extrinsic reference frames (Retz-Schmidt, 1988). The relations computed in this representation are induced by partitions of 3-D space relative to the RO. For verbal communication, reference frames that depend on the current communicational situation are needed. Hence, in a second step reference frames are superimposed to dynamically derive meaning definitions for prepositions with respect to ROs. This allows a graduated judgment of the applicability of prepositions for LO-RO pairs.

Because the intermediate representation captures the physical object constellation in 3-D space independently of reference frames, it remains *stable* even though the speaker might talk about the scene from different vantage points or may change from deictic to intrinsic reference during the communication. When his or her position changes, the speaker may use different prepositions to refer to the same portion of space. Also, the same preposition might be used with different meaning referring to different portions of space. The intermediate layer supports on a symbolic level the detection of overlapping or deviating meaning for prepositions. Only minor updating of the reference-independent representation is required from image frame to image frame if only a few objects in the scene move at the same time.

Although we are aware of the complexity of human preposition usage, we concentrate on the simple but still interesting case of developing a computational model for the prepositions left, right, in-front, behind, above, and below currently restricted to objects that are represented by convex solids.

2. OBJECT DESCRIPTIONS FROM 2-D IMAGES

2.1. A Priori Object Model

Our objects are taken from a children's toolbox, so we have rigid objects with known size and shape. This allows us to use simple geometric models (Flynn & Jain, 1991) to characterize the objects as compositions of vertices, edges, ellipses, and cylinders. The geometric models are used for reliable object recognition and 3-D scene reconstruction. The goal is to extract qualitative spatial information from real images. The geometric object models are more detailed than needed to determine spatial relations. Therefore, the objects are approximated by boxes. Similar to Abella and Kender's (1993) approach for the formulation of the 2-D relation *aligned,* an object's bounding box B_O is given as the bounding cuboid collinear to the object's principal axes.

2.2. Model-Based 3-D Reconstruction

The object recognition process starts as a knowledge-based search in regions of color-segmented images and in image contours. Additional information about the object or its location is used if it is available. Two-dimensional object hypotheses are generated. From these object hypotheses we extract model-based corrected vertices, line segments, and ellipses. A stereo matching is performed on the corrected image features using the information from the object hypotheses. Simultaneously, the best viewpoint and the object pose parameters are estimated, fitting the projection of all 3-D object models to detected 2-D image features (Socher, Merz, & Posch, 1995). The object models are fitted iteratively. The residual error decides whether a hypothesis is accepted or not. Figure 6.1 shows the image of a scene and its 3-D reconstruction. As a result, the 3-D pose of each object is computed.

3. GENERATING PREPOSITIONS FROM 3-D OBJECTS

In this section, we assume that the reconstruction of the 3-D scene is already computed. We address the problem of generating and judging the existing prepositions between located objects (LOs) and reference objects (ROs). We take into account that the meaning of prepositions depends on an underlying reference frame that in turn is determined by the actual communicational situation. Reference frames centered in the reference object, the speaker, or either some third object or person are called intrinsic, deictic, or extrinsic, respectively (Retz-Schmidt, 1988).

Instead of directly calculating prepositions from 3-D descriptions of two objects and a given reference frame (Abella & Kender, 1993; Gapp, 1994), we first construct an object-specific reference-independent spatial representation of the object constellation. This is based on acceptance relations that are induced by acceptance volumes partitioning the 3-D space in an object-specific way. In a second step, reference frames are used to compute meaning definitions for prepositions as disjunctions of acceptance relations. These meaning definitions are then applied to gradually judge the applicability of the prepositions.

3.1. Object-Specific Acceptance Relations

Let $OBJECTS$ denote the set of objects in the scene. Each object $O \in OBJECTS$ partitions the 3-D space in its own way. Based on the object's bounding box B_O the space is divided into 3-D acceptance volumes AV_i^O ($0 \leq i \leq n$, $n \in IN$). In our current approach we use 79 acceptance volumes for each object (see Fig. 6.2): The box itself forms an acceptance volume; there is one volume bound to each side of the box, two volumes bound to each edge, and six volumes bound to each vertex. A minor change of the position of another object P relative to O will effect the relationship between P and O much more when P's position is close to an edge than if it is close to the box sides. The same is true for vertices with respect to edges. This partitioning of the space is a 3-D extension of the partitioning Hernández (1993) used for his definition of object-specific orientation relations of granularity 0 and 1.

A direction vector $\mathbf{d}(AV_i^O)$ is associated with each acceptance volume. It is an approximation of the direction to which the side (edge, vertex) corresponding to AV_i^O faces in space. The vectors we currently use are shown in Fig. 6.2.

Based on the acceptance volumes AV_i^O a set of binary object-specific acceptance relations r_i^O is defined: $r_i^O \subseteq OBJECTS \times \{O\}$ with $(P,O) \in r_i^O \Leftrightarrow B_P \cap AV_i^O \neq \emptyset$. Furthermore, for each $P \in OBJECTS$ and acceptance relation r_i^O a degree of containment $\gamma(P, r_i^O) \in [0,1]$ is defined as

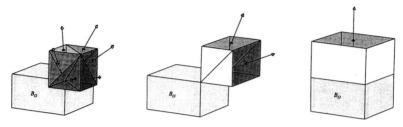

FIG. 6.2. Infinite 3-D acceptance volumes attached to an object's bounding box B_O: (a) the six acceptance volumes bound to a vertex, (b) the two acceptance volumes at an edge, and (c) the acceptance volume defined by the top side of a bounding box.

$$\gamma(P,r_i^O) = \frac{vol(B_P \cap AV_i^O)}{vol(B_P)} \qquad (1)$$

which represents the relative part of the volume of object P lying in acceptance volume AV_i^O.

3.2. The Reference-Independent Spatial Representation Layer

We have now introduced all prerequisites for representing a spatial constellation of objects independently of reference frames usually applied in human communication. This representation is called the reference-independent representation. It is stored as a relational network. Nodes correspond to objects. Each arc from some object P to another object O is labeled with pairs $(r_i^O : \gamma(P,r_i^O))$ for all those acceptance relations of O that P fulfills (i.e., $\gamma(P,r_i^O) > 0$). Figure 6.3b shows an example. For better visibility, we take a 2-D constellation. Figure 6.3b shows that object B lies with 93% of its area in AV_4^A and with 7% in AV_5^A. The reference-independent representation explicitly reflects the physical constellation of the objects in 3-D. Because usually only a few objects move at the same time only a few arcs must be updated from frame to frame. Changes of reference frames yield no changes of this representation.

3.3. Meaning Definitions of Spatial Prepositions

Reference Frames. In a 3-D environment a given reference frame *ref* is defined by three distinct axes: the front-back axis, the left-right axis, and the bottom-top axis. We represent each axis by a pair of reference vectors that are inverse to each other. The front-back axis is given by the vectors **fb** and **bf** pointing from front to back and back to front, respectively. Analo-

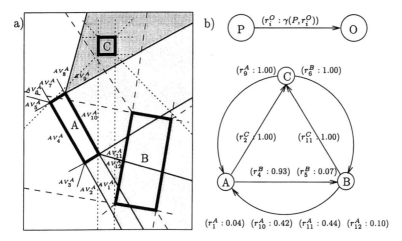

FIG. 6.3.   (a) Object-specific partitioning of 2-D space using 13 acceptance volumes; for A those volumes are shaded that intersect with other objects. (b) Reference-independent spatial representation for the 2-D scene in (a); arcs are labeled with pairs of acceptance relations and the corresponding degrees of containment.

gously, the left-right axis and bottom-top axis are represented by the reference vectors **lr, rl,** and **tb, bt,** respectively.

In the case of deictic reference—and this is what we actually apply because our objects have no intrinsic orientation—the front-back axis is calculated dynamically as the line connecting the speaker's vantage point to the reference object's center of mass. The bottom-top axis coincides with the speaker's vertical axis, and the left-right axis is defined to be orthogonal to both other axes. In the case of an intrinsic reference system, the vectors should be known a priori. For extrinsic reference frames other context information, for example, the current direction of movement of an object, might be used.

Determination of Meaning Definitions.   We define spatial prepositions as disjunctions of acceptance relations. A labeling procedure determines for each reference object RO, reference frame *ref*, and preposition *prep* which set of RO's acceptance relations captures the meaning of *prep*. This set is denoted as *def(ref,prep,RO)*. Each acceptance relation r_i^{RO} chosen for the definition is associated with a degree of accordance $\alpha(ref,prep,r_i^{RO})$ that expresses how strong it is compatible with the meaning of *prep*.

For example, we consider the labeling for the preposition in-front. The reference vector **bf** of the current reference frame is taken to determine the definition set of in-front. An acceptance relation r_i^{RO} is incorporated in *def(ref,*in-front,RO) if the inner product $\langle \mathbf{d}(AV_i^{RO}) \mid \mathbf{bf} \rangle > 0$. Its degree of accordance is computed as

$$\alpha(ref, \text{in-front}, r_i^{RO}) = 1 - 2 \cdot \frac{\arccos(\langle \mathbf{d}(AV_i^{RO}) \mid \mathbf{bf} \rangle)}{\pi}. \tag{2}$$

This labeling is motivated by the following observation used in computer graphics: A surface point of a convex object with a continuous surface function is visible if the inner angle of its normal vector $\mathbf{n}$ and the vector $\mathbf{v}$ pointing from the surface point to the observer's vantage point is less than 90°. The aim of our labeling for the relation in-front is to gather all those acceptance relations that correspond to those object sides, edges, and vertices that are visible from the speaker's vantage point. However, our objects are abstracted by bounding boxes that have no continuous surface. Hence, we adapt and simplify the criterion by taking the direction vectors $\mathbf{d}(AV_i^{RO})$ instead of the surface normals and by taking the reference vector $\mathbf{bf}$ instead of $\mathbf{v}$.

The degree of accordance linearly decreases with an increasing inner angle between $\mathbf{d}(AV_i^{RO})$ and $\mathbf{bf}$. That is, the stronger an acceptance volume faces the vantage point, the larger the degree of accordance. At the current stage this yields satisfying results. However, the effects of nonlinear functions that prefer small inner angles need more attention.

The labeling for behind, right, left, above, and below is calculated similarly. For the determination of their definition sets the reference vectors $\mathbf{fb}$, $\mathbf{lr}$, $\mathbf{rl}$, $\mathbf{bt}$, and $\mathbf{tb}$ are chosen, respectively. Figure 6.4 shows examples of the determination of definition sets for a 2-D scene. The cases (a) and (b) where the reference vectors are collinear with the edges of the reference object's bounding boxes show that our labeling yields the results intuitively expected.

The symbolic definitions of prepositions have another advantage. They allow symbolic testing for overlapping or shared meaning of the prepositions used in the verbal communication. This is interesting if the speaker changes his or her position during the communication and refers to the same physical spatial configuration using different prepositions. The prepositions behind in Fig. 6.4 (a) and (c) have similar but different meaning. They share four acceptance volumes.

3.4. Generating Spatial Prepositions

Based on this dynamically calculated definition, the applicability of a preposition $prep(ref, \text{LO}, \text{RO})$ for a given located object LO with respect to a reference object RO is determined in the following way: If $def(ref, prep, \text{RO})$ is contained in the label set of the arc from LO to RO in the reference-independent spatial representation then $prep(ref, \text{LO}, \text{RO})$ is applicable. Its degree of applicability $\delta(ref, prep, \text{LO}, \text{RO})$ is calculated as follows:

$$\delta(ref, prep, \text{LO}, \text{RO}) = \sum_{r_i^{RO} \in def(ref, prep, \text{RO})} \alpha(ref, prep, r_i^{RO}) \cdot \gamma(\text{LO}, r_i^{RO}). \tag{3}$$

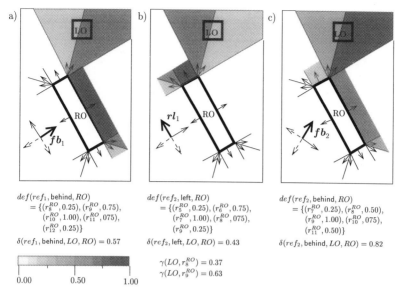

$def(ref_1, \text{behind}, RO)$
$= \{(r_8^{RO}, 0.25), (r_9^{RO}, 0.75),$
$(r_{10}^{RO}, 1.00), (r_{11}^{RO}, 075),$
$(r_{12}^{RO}, 0.25)\}$
$\delta(ref_1, \text{behind}, LO, RO) = 0.57$

$def(ref_2, \text{left}, RO)$
$= \{(r_5^{RO}, 0.25), (r_6^{RO}, 0.75),$
$(r_7^{RO}, 1.00), (r_8^{RO}, 075),$
$(r_9^{RO}, 0.25)\}$
$\delta(ref_2, \text{left}, LO, RO) = 0.43$

$def(ref_2, \text{behind}, RO)$
$= \{(r_7^{RO}, 0.25), (r_8^{RO}, 0.50),$
$(r_9^{RO}, 1.00), (r_{10}^{RO}, 075),$
$(r_{11}^{RO}, 0.50)\}$
$\delta(ref_2, \text{behind}, LO, RO) = 0.82$

0.00   0.50   1.00

$\gamma(LO, r_8^{RO}) = 0.37$
$\gamma(LO, r_9^{RO}) = 0.63$

FIG. 6.4. Meaning definitions and degrees of applicability for prepositions behind and left and different reference frames. A 2-D scene is chosen for better visibility. RO corresponds to A in Fig. 6.3. In all illustrations the acceptance volumes chosen for the prepositions definition set are shaded. The darker the shading the higher the degree of accordance of the corresponding acceptance relation. The reference frames are indicated by the coordinate system in the lower left.

Figure 6.4 shows the degrees of applicability for the prepositions behind and left for different reference frames in 2-D.

4. USING PREPOSITIONS FOR OBJECT LOCALIZATION IN 2-D IMAGES

Now, consider an utterance such as *"the rhomb-nut right of the tire."* The preposition *(right)* referring to the reference object *(tire)* is used to find the located object *(rhomb-nut)* in the listener's visual space. In this section, we show how to reduce the recognition effort for the located object.

Assume that the 3-D location of RO is computed. Suppose that for a given preposition *prep* and a reference frame *ref* the meaning definition $def(ref,prep,RO)$ is also given. The definition is used to derive the LO's admissible image regions. For each acceptance relation $r_i^{RO} \in def(ref,prep,RO)$ with a degree of accordance α above some threshold $\in [0,1]$ the corresponding acceptance volume AV_i^{RO} is projected onto the image. All resulting polygons are merged to yield that image region in which the LO may be located. Figure 6.5 shows the projected acceptance

FIG. 6.5. Projected acceptance volumes AV_i^2 corresponding to relations $r_i^2 \in def(ref,\text{right},2)$ with $\alpha(ref,\text{right},r_i^2) > .3$; the maximum value for α is found in the darkest region. The objects are numbered as in Fig. 6.1.

volumes (darker regions) for the preposition right with respect to the socket (2). The threshold for α is set to .3.

The threshold prevents the projection of those acceptance volumes onto the image that weakly coincide with the preposition's meaning but may yield large image regions. It also introduces another possibility: If hints from the communicational context are available that express whether a preposition is to be interpreted in a stronger or a weaker sense the threshold could be dynamically determined. However, this is an issue we have to work on in the future.

5. RESULTS

Various experiments have been performed with real image data. So far, we have concentrated on the evaluation of the qualitative results we obtain for natural scenes with multiple objects, and on testing the behavior of the degree of applicability δ for the prepositions.

As an example we present the spatial prepositions for the numbered objects in the scene shown in Fig. 6.1. The chosen reference frame takes the position of one camera as vantage point to allow an easy verification of the qualitative results in the images. Each of the five objects functions as LO and as RO. The degrees of applicability of all six prepositions for every object pair are presented in Table 6.1. The value of the maximally judged preposition for an object pair is printed in bold type. The results

TABLE 6.1
Degrees of Applicability

LO	RO	left	right	above	below	behind	front	LO	RO	left	right	above	below	behind	front
2	1	0	0.20	0.20	0	0.69	0	1	2	0.29	0.15	0	0.04	0	0.53
3	1	0	0.20	0.17	0	0.71	0	1	3	0.45	0.08	0	0.06	0.03	0.41
4	1	0.21	0	0.20	0	0	0.69	1	4	0.01	0.55	0	0.11	0.40	0
5	1	0.20	0	0	0.18	0	0.70	1	5	0.17	0.31	0.02	0	0.50	0
3	2	0	0.62	0	0	0	0.38	2	3	0.60	0	0.10	0.01	0.36	0
4	2	0.37	0	0	0	0	0.63	2	4	0	0.39	0.11	0	0.57	0
5	2	0.12	0.02	0	0.13	0	0.78	2	5	0	0.17	0.14	0	0.75	0
4	3	0.62	0	0	0	0	0.38	3	4	0	0.68	0.13	0.01	0.25	0
5	3	0.25	0	0	0.17	0	0.66	3	5	0	0.39	0.17	0	0.52	0
5	4	0	0.67	0	0.26	0.20	0.01	4	5	0.60	0	0.22	0	0.14	0.18

Note. The degrees of applicability are calculated for the six prepositions for the objects 1 to 5 shown in Fig. 6.1. Each object functions as reference object and as located object.

are good and correspond to our expectations. A very good example is the object pair 5-2. The tire (5) is located in front of the socket (2). It is not exactly in front but slightly left as can be seen in the results. In this case, the spatial location is invertible. The tire lies in front and the socket behind the tire with a slight right tendency. The spatial configuration for the long wooden bar (1) and the short wooden bar (4) is not invertible. Because of the difference in size, it is sufficient to say: The short bar (LO = 4) is in front of the long one (RO = 1). To qualitatively locate object 1 with respect to object 4, that is, LO = 1 and RO = 4, two prepositions must be used: The long bar (1) lies right and behind the short bar (4). This is very well detected.

The relative size of the objects has another surprising consequence. The tire (5) is judged as being in front *and* left of the long wooden bar (1). This results from the fact that the acceptance relation associated to the long side of the bar is contained in the definitions for the prepositions front *and* left. $\alpha(ref,\text{left},r_i^1)$ is quite small but because the whole tire is contained in this acceptance volume, the weighting causes a degree of applicability of $\delta(ref,\text{left},5,1) = .2$.

We are dealing with real and therefore noisy data. Hence, the 3-D scene reconstruction is slightly erroneous. In particular, for our example scene the reconstruction process does not place all objects on the same plane, even though they were presented to the stereo camera on a table. This explains the fact that we get positive degrees of applicability for the prepositions above and below. Further work will include how to cope with noisy results.

The degree of applicability δ depends on the location of LO and RO as well as the reference frame and the preposition. For example, we consider the behavior of δ as a function of the location of LO with respect to RO for the prepositions left and right. The reference frame is chosen as previously and the located object is rotated on a full circle in the plane given by lr and fb. The radius of the rotation circle is the distance between LO and RO. Figure 6.6 shows the curve for $\delta(ref,\text{left},1,4)$ and $\delta(ref,\text{right},1,4)$. As expected the curves for the two antagonistic prepositions behave in opposition to each other. The maximum degree is at about .7. None of the $d(AV_i^1)$ is parallel to rl or lr, therefore none of the acceptance relations gets the maximum degree of accordance.

6. DISCUSSION

In the discussion of our approach with respect to other work in the field we concentrate on object abstraction, reference-independent spatial representation as an intermediate representation layer, judgment of prepo-

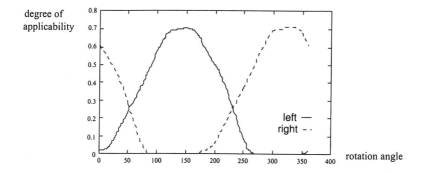

FIG. 6.6. Behavior of δ(*ref*,left,1,4) and δ(*ref*,right,1,4) when a 360° rotation of the located object (LO = 1) around the reference object (RO = 4) is simulated. The reference frame remains stable.

sitions, handling of reference frames, and the usage of prepositions for the restriction of possible image positions of objects.

Abella and Kender (1993) took bounding boxes collinear to the image axes for calculating the relations *above* and *below*. For our needs this object abstraction is too coarse. Hence, our object abstraction is a 3-D adaptation of Abella and Kender's modeling of the 2-D relation *aligned*. It is a better abstraction from the actual object's shape because it captures the main directions of the object's extension. It is important to note that the principal structuring of our model could just as well be based on any other object abstraction.

One major difference to Abella and Kender (1993), André et al. (1988), and Gapp (1994) is our introduction of an intermediate reference-independent representation layer that relationally expresses the physical object constellation in 3-D space. This facilitates dealing with image sequences of scenes where usually only few objects change positions at the same time. Hence, although it has to be detected whether objects have moved or not within two sample time points, only the acceptance relations for objects that changed position have to be recalculated. The idea to introduce an intermediate object-centered representation layer that is based on meaningless relations with respect to the highest interpretation level was mainly influenced by the work of Hernández (1993). In his purely qualitative 2-D modeling approach he used an analogical object-centered representation structure. Orientation relations of different levels of granularity are provided (together with topological relations) to express the relative 2-D position between objects. In the realization of this model a set of 20 "abstract" qualitative relations is used to model different sets of orientation relations with respect to different levels of granularity. Each higher level relation is defined a priori as some specific disjunction of these abstract relations. During constraint propagation, this "common grounding" allows one to detect overlapping or contradicting meaning of relations of different granularity. In our approach we adapted this idea. Prepositions are defined as disjunctions of acceptance relations. However, because we have to deal with changing positions of objects and/or communicators in a numerically described environment, the meaning of the acceptance relations is not fixed until the objects are placed in 3-D space. Furthermore, the meaning definitions of the prepositions are dynamically computed.

Abella and Kender (1993), André et al. (1988), Gapp (1994), and Wazinski (1993) correctly emphasized that spatial relations must be regarded as being fuzzy or graduated. Gapp reflected the fact that the extension of the objects to be related influences the usage of prepositions by humans. Therefore, he calculated the angle between the centers of mass of the located object and the reference object with respect to a specific coordinate system reflecting size and extension of the reference object. He also included the distance between the objects to calculate the degree of applicability of a preposition. In our case, looking only for the placement of the located object's center of mass with respect to the acceptance volumes of the reference object would consequently lead to nonfuzzy preposition definitions. We solve this problem by using the LO's bounding box instead of its center of mass. By placing the located object's bounding box in the space partitioned by the reference object we account for both demands mentioned. Our object-specific partitioning of the 3-D space counterparts Gapp's object-specific coordinate system because the size of the reference object influences the size of its acceptance volumes. That is, our model accounts for the relative size between LO-RO pairs. We have not yet incorporated that the degree of applicability of prepositions decreases with an increasing distance between objects.

Several systems (André et al., 1988; Gapp, 1994; Hernández, 1993) have accounted for the fact that spatial models that support man-machine communication must be able to deal with different reference frames to allow for adequate communication with a human partner (Herrmann, 1990; Retz-Schmidt, 1988). As shown earlier in the chapter, we can handle different reference frames. Additionally, our modeling approach enables the efficient detection of overlapping or deviating meaning of preposition relations that are constructed on the basis of different reference frames.

The aspect of object identification on the basis of given spatial relations was discussed by Gapp (1994). There are major differences between his and our approach. He assumed that the image interpretation works unidirectionally. First, all objects in the image are recognized and their 3-D positions are calculated. Based on this knowledge higher level interpretations are derived. Object identification in this context means to efficiently look for an object description that has already been computed by the low-level recognition. On the opposite, we envision a system where a bidirectional control flow between lower level recognition tasks and higher level interpretation tasks is possible. That is, information may pass from higher levels to the lower level processes. Consequently, our understanding of object identification means the ability to derive expectations regarding the possible position of an object in an image. We have shown in section 4 how this expectation is computed. The benefit we expect from this approach is that the low-level efforts can be much better tailored to what is actually needed by the higher level interpretation processes.

7. CONCLUSION AND OPEN PROBLEMS

We introduced a spatial 3-D model using the prepositions left, right, in-front, behind, above, and below for objects represented as convex boxes. The major contribution of our computational model is that it is suitable for the generation of prepositions from numerical data and, in opposite to other approaches, it also allows one to use spatial prepositions to restrict the expected image position of located objects and hence the low-level analysis can benefit from high-level qualitative knowledge. The model accounts for specific demands resulting from a man-machine interaction scenario where the communication takes place via spoken language and the communication partners may change their positions over time: It can handle different types of, and dynamically changing, reference frames; it supports the testing for overlapping or deviating meaning of prepositions. Furthermore, by introducing an intermediate representation layer independent of actual reference frames image sequences can be handled efficiently.

In the future, we plan to investigate the cognitive adequacy of our realization of the prepositions through psycholinguistic experiments. Depending on their results, we may adapt our judgment of the prepositions or may even adapt the granularity of the partitioning of 3-D space currently chosen. In particular, we wish to use the experiments to better understand how the relative distance and size of objects influence the use of prepositions. As a consequence, we will incorporate this understanding in a more adequate formulation of the judgment for the degree of applicability.

ACKNOWLEDGMENTS

This work has been supported by the German Research Foundation (DFG) in the project SFB 360.

REFERENCES

Abella, A., & Kender, J. (1993). Qualitatively describing objects using spatial prepositions. In *Proceedings of AAAI-93* (pp. 536–540). Cambridge, MA: MIT Press.

André, E., Herzog, G., & Rist, T. (1988). On the simultaneous interpretation of real world image sequences and their natural language description: The system SOCCER. In Y. Kodratoff (Ed.), *Proceedings of the 8th ECAI* (pp. 449–454). London: Pitman.

Cui, Z., Cohn, A., & Randell, D. (1992). Qualitative simulation based on a logical formalism of space and time. In *Proceedings of AAAI-92* (pp. 679–684). Menlo Park, CA: AAAI Press.

Flynn, P. J., & Jain, A. K. (1991). CAD-based computer vision: From CAD models to relational graphs. *IEEE Trans. Pattern Analysis and Machine Intelligence PAMI, 13*(2), 114–132.

Gapp, K.-P. (1994). Basic meanings of spatial relations: Computation and evaluation in 3D space. In *Proceedings of AAAI-94* (pp. 1393–1398). Cambridge, MA: The MIT Press.

Guesgen, H. W. (1989, August). *Spatial reasoning based on Allen's temporal logic* (Tech. Rep. No. TR-89-049). Berkeley, CA: International Computer Science Institute.

Hernàndez, D. (1993). *Qualitative representation of spatial knowledge* (Lecture notes in artificial intelligence, 804). New York: Springer-Verlag.

Herrmann, T. (1990). Vor, hinter, rechts und links: Das 6H-Modell [In front, behind, right, and left: The 6H model]. *Zeitschrift für Literaturwissenschaft und Linguistik, 78*, 117–140.

Herskovits, A. (1986). *Language and spatial cognition.* Cambridge, MA: Cambridge University Press.

Olivier, P., & Tsujii, J. (1994). *Prepositional semantics in the WIP system.* AAAI-94 Workshop on Integration of Natural Language and Vision Processing [workshop notes].

Retz-Schmidt, G. (1988). Various views on spatial prepositions. *AI Magazine 9*(2), 95–105.

Schirra, J. R. J., & Stopp, E. (1993). ANTLIMA—A listener model with mental images. In *Proceedings of 13th IJCAI* (pp. 175–180). San Mateo, CA: Morgan Kaufmann.

Socher, G., Merz, T., & Posch, S. (1995). 3-D reconstruction and camera calibration from images with known objects. In D. Pycock (Ed.), *Proceedings of the British Machine Vision Conference* (pp. 167–176). Birmingham, England: Breva Press.

Stopp, E., Gapp, K.-P., Herzog, G., Laengle, T., & Lueth, T. C. (1994). Utilizing spatial relations for natural language access to an autonomous mobile robot. In *KI-94: Advances in Artificial Intelligence, Proceedings of the 18th German Annual Conference on AI* (Lecture notes in artificial intelligence, 861, pp. 130–141). Berlin: Springer.

Wazinski, P. (1993, May). *Graduated topological relations* (Tech. Rep. No. 54). Saarbrücken, Germany: Universität des Saarlandes.

7

Shapes From Natural Language in VerbalImage

P. Bryan Heidorn
University of Illinois, Urbana–Champaign

The integration of computational models of vision and natural language processing has significant practical and theoretical consequences. This chapter describes an information retrieval task where this integration is crucial. Language can be used to describe the environment. These descriptions can be broken into two classes: one class describing the location of objects and the other describing the objects themselves. VerbalImage is a computational implementation of a theory of the description of objects and in particular the shape of objects. A small number of words are identified that denote the likewise small number of prototypical shapes in a domain. A closed class of words and phrases signals shape modification of these prototypes. As suggested by prior theories of the semantics of shape modification, words and phrases such as *thin* or *1 inch long* are sensitive to abstracted or idealized features of the shape prototypes of the objects of the domain. These abstract features of prototype shapes are limited in number and include in addition to other features primary and secondary dimensionality, and intrinsic "up," "front," "left," and "right." In VerbalImage, natural language (NL) descriptions of physical objects are interpreted and rendered on a graphic screen. Additional NL descriptions can be used to modify the image on the graphics screen. The main contribution of this work is to show that linguistic and perceptual theories typified by Jackendoff (1991) and Biederman (1987), when appropriately modified, are useful in a computational setting. There are also new contributions in the area of subclassification.

1. INTRODUCTION

The evolutionary advantage granted to humans by the ability to verbally describe their environment is obvious. Humans may communicate knowledge about their environment to others in the absence of the original environment. Of particular interest in this chapter is the ability to describe objects.

A listener is able to visualize nonpresent objects and recognize objects in the class when they are later encountered. Landau and Jackendoff (1993) argued that the human spatial representation system is relatively rich in its ability to describe shape but relatively limited in the ability to use object shape to specify spatial relations. They reviewed literature to support the assertion that there are separate "what" and "where" channels and that these are somewhat localized in the human brain. In their *Design of Spatial Representation Hypothesis*, Landau and Jackendoff stated that the differences in language systems between object identification (nouns) and object localization (spatial prepositions) is attributable to the underlying organization of "what" (identification) and "where" (localization) channels. This chapter explores the use of language to describe the shape of objects, that is, the description of "what." Rather than concentrate on proper names as is done in Landau and Jackendoff's treatment, the current focus is on shape prototype specification and modification.

For the purposes of this work it is useful to classify identification tasks as superordinate, basic, and subordinate (Rosch & Mervis, 1975). Hoffman and Richards (1984) argued that people recognize objects by parts at the basic level. Yet, at the subcategorization level, objects share parts (Tversky & Hemenway, 1984). Embarking from this point the current work builds on the inference that the distinctions between object subcategories are based on fine distinctions between part parameters and relations. Whereas class prototypes are all about similarity, subcategorization is all about differences and these differences are most easily described as differences from a prototype. Biederman (1987) proposed a small set of such parameters. Quantitative versions of these are used for image indexing in VerbalImage.[1]

2. SHAPE LANGUAGE

Different word classes and syntactic modification structures serve distinct roles in the specification of shape. The semantics of these words is not

---

[1]The relationship between the names of "Spoken Image" (Ó Nualláin, Farley, & Smith, 1994) and "VerbalImage" is purely coincidental. Spoken image addresses the spatial relations of objects whereas VerbalImage addresses shape. VerbalImage is a name adopted in the dissertation proposal of the author in the spring of 1994 and Spoken Image was published first. Because there is a complementary relationship between the packages, the original name has been maintained.

determined by the shape of the objects being described by language but rather on a finite set of abstract features or parameterizations of shape. A related schematic approach to spatial prepositions is proposed by Herskovits (1985, 1986). Different word classes are sensitive to variation along different parameter values. The following parameters have proven to be useful in the implementation of the shape-based retrieval system that is described later in this section. The parameters include primary and secondary dimensionality, prime dimensions of height, width, and depth, and cross-sectional symmetry. The features of cross-sectional curvature and axis curvature are noninformative in the domain. They do not differentiate objects in the domain. The meanings of these terms are generally obvious for this discussion. The use of these features was inspired primarily by work of Jackendoff (1991) and Biederman (1987). The main word classes of interest here are object class names, marked and unmarked nominal shape adjectives, and dimension-sensitive shape adjectives.

Within a class of objects that may be identified by shape, there are a set of object class names that signify the classes. Terms such as *leaf* or *acorn* denote a class of objects (or parts of objects) that share gross shape properties. Acorns share the properties of three-dimensional (3-D) primary dimensionality, one-dimensional (1-D) secondary dimensionality, cross-sectional reflective and rotational symmetry, curved cross section, generating axis expansion coefficient, and uncurved generating axis. The use of the term *acorn* in an utterance in effect signifies the values for all of these properties.

Nominal shape adjectives may be marked or unmarked depending on the typicality of their use as shape indicators for the object class. Any object with a fixed shape can be used to impose shape on the head noun. Some of these nominal adjectives have privileged status and may signify cognitive reference points or prototypes for the class as discussed later. These include shape terms such as *oval*, *round*, and *square*. When used in a structure like "oval leaf," they imbibe or constrain their shape to the head noun. They do not, however, change the value of major shape parameters that define the head noun's object class. In this domain the protected shape properties are those specified in the previous paragraph. Consequently the meaning of these terms varies with the shape properties of the objects they modify. An "oval leaf" retains the primary two-dimensional (2-D) dimensionality, so is still "flat," and an "oval acorn" keeps a primary 3-D dimensionality. In the case of the leaf and acorn, the term *oval* causes a change in the generating axis expansion coefficient. It specifies how quickly the object tapers toward the ends.

Nominal shape adjectives are marked when it may not be obvious to the listener that the shape parameters are the important characteristic of the modification (as opposed to color, texture, or other property), thus

"egg-shaped acorn." Again, major shape abstractions such as dimensionality and cross-sectional symmetry are not modified by the construct. An
"egg-shaped leaf" does not have cross-sectional rotational symmetry.

There are a small set of shape-modifying adjectives that act directly on
particular parameters. The meaning of these terms is often dependent on
the object class–specified shape parameter values of the objects they
modify. VerbalImage implements semantics for dimensionality sensitive
modifiers. The meaning of words and phrases like *wide(er)*, *long(er)*, and
1.2 inches long is dependent on the primary dimensionality. The meaning
of "wide" in "a wide, oval acorn" is that the width[2] and depth are greater
than usual by some function. In VerbalImage the modification is arbitrarily set to 20%. The meaning of "a wide, oval leaf" does not involve the
modification of the thickness of the leaf. Note that "wide" overlooks the
default dimensionality of "oval" but is sensitive to the dimensionality of
the head noun "leaf."

Objects with shape may be viewed as residing in a low-dimensional
shape space. The dimensions of the space are the abstract shape parameters listed earlier. Object class names like *leaf* and *acorn* signify regions of
the space where parameter values are set to match the members of the
class. For the purposes of this chapter, a shape prototype is a region in
the shape space of a class of physical objects capable of being referenced
linguistically. The prototypes region in the space signifies a set of shape
parameters that differentiate it from other class members. By definition,
members of a grouped class of objects are more alike than different.
Prototypes are selected to significant or salient differences between members of the class for subcategorization purposes as is done in the information retrieval task discussed later. Prototypes serve as cognitive reference points that signify collections of parameter values. Prototypes need
not be exemplars because they need not refer to an extant object.

3. IMPLEMENTATION

In both the linguistic and visualization aspects of VerbalImage, a compo-
nential representation of objects is used. This is consistent with many
perception theories (Biederman, 1987; Hoffman & Richards, 1984; Marr,
1982). Objects are composed of collections of relatively simple parts. The
richness of visual complexity arises from a combination of low-dimen-

---

[2]Unfortunately, as pointed out here the meaning of "width," "length," and "depth" is
relative. In this chapter, the main generating axis is the length (Y axis). The width specifies
a distance from right to left (X axis) and depth is the remaining orthogonal axis (Z axis,
projecting into the paper).

sional parameterization of parts and the relationship between multiple parts. There is a long tradition of describing object parts as idealized forms, including among others Marr's generalized cones and Biederman's Geons. These approaches are geared toward human recognition of basic-level categories. This level of analysis allows for recognition of distinctions between objects such as chairs and tables. Because of the requirements of the subcategorization task, VerbalImage uses the more detailed quantitative representation for indexing. This contrasts with language semantics that are more idealized and use qualitative features.

The corpus of text used in this study is the descriptions of 40 species of oak trees provided in a layperson field guide to tree identification (Brockman, 1968).

The following example is typical of the corpus:

Red Oak Group: Northern Red Oak *(Quercus rudra)* has deciduous leaves, 5 to 8 inches long and 4 to 5 inches wide, with 7 to 11 pointed toothed lobes separated by regular sinuses that extend halfway to the midrib. The leaves turn red in the fall. The oblong-ovoid acorns are 0.8 to 1 inch long, with a flat, saucer-like cup at their base. The dark-brown to black bark is ridged and furrowed. Grows 50 to 70 feet tall and 1 to 3 feet in diameter, with a rounded crown. (Brockman, 1968, p. 126)

VerbalImage processes the sentences that follow to produce the images in Fig. 7.1.

1a)   The acorn is roundish.

1b)   The acorn is 1.2 inches long.

1c)   The acorn is wider.

1d)   The acorn is wider.

The focus of the current work is on shape. Given this constraint, no attempt is made to address the issues of color, texture, amorphous objects, motion, the spatial relation between objects, or the location on objects in space. Later versions of the program will be expanded to address these issues.

The next sections describe the main components of the system, concentrating on the domain model and language processing.

3.1. VerbalImage Architecture

VerbalImage is an image retrieval system that accepts natural language descriptions of objects, generates images of those descriptions, and finally retrieves images that are "similar" to the described image. The work

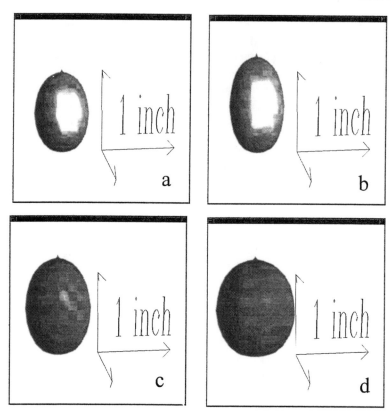

FIG. 7.1. Declarative shape modification.

described here is a type of object identification and subcategorization task, identifying particular types of known object classes. The test domain is tree identification based on leaf and seed shape description.

A goal of VerbalImage is to generate a detailed machine representation of an object that a human wishes to identify. The machine representation is in a form that allows for retrieval of similar objects, with similarity defined over the shape, not the words used to describe the shape. This is accomplished by establishing a verbal-to-visual feedback loop between the user and the system. The use of a feedback loop presupposes that the user has a conceptual representation of the object of interest as well as a facility to measure the difference between this internal representation and a visual representation of the object. The user is able to articulate this difference verbally. VerbalImage accepts a verbal description of an object and uses domain knowledge to interpret the description and generate an object-based model of the object in question. This symbolic object representation can reexpress itself as a graphic on a computer screen. Subsequent descrip-

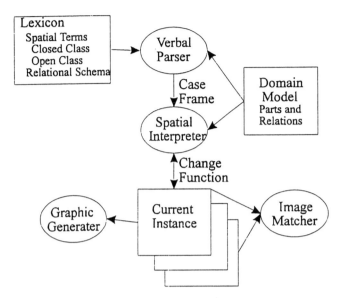

FIG. 7.2. VerbalImage architecture.

tions by the user serve to modify the internal parametric settings of the computer model and subsequently the graphic screen representation.

The main components of VerbalImage that relate to linguistics are depicted in Fig. 7.2. The main components are a part-wise hierarchical domain model, D (as depicted in Fig. 7.3); a natural language processor, S, that maps linguistic strings into instantiated components of the domain model; a graphics mapper, G, that transforms the parameterized domain model into graphic images; image analysis routines, I, and a matcher. In VerbalImage, parts are represented as named prototypes. The graphic forms of these prototypes are represented as polyhedrons. It is important to note that the actual form of the representation for the graphic component of the system is independent of the linguistic component. The linguistic component uses particular abstract parameters, A_C of the basic shape semantics to be applied properly.

The syntactic parser in this study is a relatively simple and slightly modified version of a parser kindly provided by Jeff Siskind of the University of Toronto.[3] The output of the parse is a functionally tagged parse tree of the form (S (NP (D The) (N acorn)) (VP (BE is) (NP (NUM 1.2) (N inch) (A long))). The semantic interpreter[4] takes this tree as input

---

[3]The original parser may be found at http://www.cdf.toronto.edu/DCS/Personal/Siskind.software.hmtl

[4]Parser modifications, a grammar, and semantic interpreter are available from the author of this chapter.

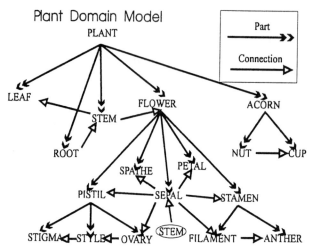

FIG. 7.3.   Part hierarchy.

and creates instances of parts of a knowledge base as output. The semantic processor also assigns values to parameters of objects in the domain. Count nouns are assumed to represent fixed-form objects or object parts, an admittedly simplistic view but appropriate for the task. Shape adjectives and phrases act to modify the parameters of these objects. Verbs describe the relations between parts.

The knowledge base for the domain is a decomposition shape hierarchy with whole objects at the root and parts as branches. Cross-links between branches describe potential relations between parts (Fig. 7.3). Physical objects represented in symbolic form can generate procedures to reexpress themselves graphically.

Objects, parts, and relations are implemented as CLOS classes. During semantic interpretation these classes are instantiated.

More formally, U is a set of shape utterances. In this case U is the set of 40 tree descriptions plus a few additional language forms. Semantic interpretation is defined as a set of functions, S. A_C is the set of shape parameters for physical objects in the domain model D. $\hat{A}$ is the subset of A_C that is relevant to lexical semantics. Some other non-$\hat{A}$ members of A_C are used for the visualization module. Potentially, other parameters are used in spatial and shape reasoning. $S_{\hat{A}}$ indicates semantic sensitivity to abstract properties of the shape (such as dimensionality). The equation $S_{\hat{A}}(U,D) \Rightarrow O_C$ states that the semantic interpretation of U, sensitive to $\hat{A}$ changes some set of attributes A_C of object O_C. G is a set of functions that map object attributes A_C into a graphic representation. O_G is a graphic object representation. $G(O_C) \Rightarrow O_G$ is the mapping of an object from the conceptual model domain to the visual domain. Finally, I is a set of functions that analyzes O_G to produce image-indexing attributes, A_I, such that $I(O)_G) \Rightarrow A_I$. S, G, and I are sets of mapping functions. O_C and O_G are objects (or parts) in the conceptual or visualization domain respectively. A_C and A_G are attributes of the associated object types and A_I may be seen as attributes of the stored image object but used in the index. A_G, A_C, and A_I are not equivalent.

In this object-oriented architecture, all objects encapsulate the procedures that transform each object type into the next. Of particular interest for language-to-visualization mapping is the identity of the object abstractions $\hat{A}$. In this work to date only primary and secondary dimensionality and intrinsic front are necessary.

The independence of representation licensed by S allow multiple synonymous expressions to describe the same physical object. G allows one unique visualization of the current system state, granting the user language-independent feedback. Image indexing and retrieval require more fine-tuned discrimination than provided to the linguistic parameters $\hat{A}$. For example, the expansion ratio of the cross section is often inherited from the object domain model without need for an implicit linguistic description.

All objects are represented in an object-centered coordinate system. This simplification eliminates the need to store orientation explicitly in the object model. This will need to be extended to include explicit orientation once interobject spatial relations are modeled.

3.2. Prototype and Shape Assignment

The next two sections address issues of the mapping of natural language to the domain model, $S_{\hat{A}}(U,D) \Rightarrow O_C$. For any class of fixed-form objects, humans tend to identify a set of typical shapes that break the class into smaller sets. The number and size of sets are determined by task parameters. It is natural for these sets to have names that specify the shape. Here we examine a class of shape adjectives that specify shape prototypes for a particular shape domain. In the test corpus, the following adjectives are used to specify acorn shape: *conical, ellipsoidal, oblong, oblong-ovoid, obovoid, oval, ovoid, ovoid-oblong, round,* and *rounded.* Each of these words in the lexicon is associated with a procedure, S, tied to physical objects. S alters the parameter settings A_C of the instances of the domain model. Some shape adjectives are directly tied to prototypes. In these cases, such as "round," the name of the prototype is added to the prototype-shape slot. Other adjectives designate modified prototypes. In these cases the associated procedure assigns a prototype name and it varies other parameters. In the case of "oblong-ovoid," the prototype is set to "ovoid" and the length parameter increased. The pure prototypes, those with polyhedral representations requiring no parametric modification, are displayed graphically in Fig. 7.4.

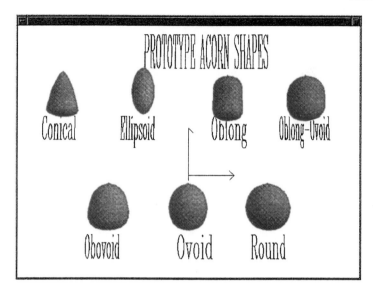

FIG. 7.4. Acorn shape prototypes.

The "oval" in an "oval acorn" is very different from an "oval leaf" in that the leaf is primarily 2-D. Yet, there is a similarity between the shapes. The semantic interpretation of the nominal adjectives is the same. The same prototype parameter values are set in the domain model. The dimensionality arises from the leaf and acorn class parameters, not from the adjective *oval*. As is seen later, the semantics of other shape-modification adjectives, such as *wide*, may depend on these dimensional features imposed by the main noun.

At the point of image rendering, the difference in primary dimensionality may also have an effect. The 2-D primary dimensionality conflicts with the 3-D dimensionality of the prototype. This leads to an idealization of the prototype form during visualization, reducing the image dimensionality by one.[5] There is one underlying prototype for an "oval," "acorn/leaf/dish/planetoid." It is object constraints like dimensionality that tune the meaning.

3.3. Shape Modification

As seen previously, some shape words imply a particular shape but other words and phrases are used to indicate modification of a shape. In the

---

[5]In VerbalImage this is accomplished by using a second prestored polyhedron. A more general and conceptually pleasing solution would be a function that performed the dimension reduction on an arbitrary shape.

current lexicon, this includes words and phrases like *wide(er)*, *thin(er)*, and *1 inch long*. Again the semantics bind the terms to procedures that change parameters in the domain model. As can be seen in the definition of *wide* that follows, some of these procedures are sensitive to the dimensional structure of the object they modify (see also Fig. 7.5).

In the sentences, "The [leaf, acorn] is .5 inches wide," the width of the leaf does not affect the thickness, but in the case of the acorn it does because of the constraints imposed by cross-sectional symmetry.

The shape-modification words *long*, *wide*, and *thick* have two forms based on their modifiers. They may specify absolute values for particular dimensions of the object as in "The X is 1.5 inches [long, wide, thick]." The associated *S* functions for this form assign the value of Y, X, and Z axis respectively to the specified value. In the relative form, "The X is (long, wide, thick)," *S* alters the axis scale as a function of the values inferred from the prototype.

These assignments may be extended to dimensions not explicitly mentioned in the description. That is, these functions, S, are sensitive to the shape abstraction parameters. This action is triggered by the parameters Â. These parameters include dimensionality and symmetry. For example, a sphere is constrained to have equal length, width, and depth. "A 1-inch-round acorn" is also 1 inch wide and deep. "A 1-inch-long oval acorn" specifies little about the exact width but "oval" must maintain an aspect

```
(defmethod DeltaWidth
    ((PObj PhysicalObjectProperties) ChangeSize &key Absolute)
(with-slots
        (PrimaryDimensionality SecondaryDimensionality
        AbsLength AbsWidth AbsDepth RelLength RelWidth RelDepth) PObj
    (if absolute
    (cond
        ((eq PrimaryDimensionality 3)
        (setf AbsWidth Absolute)
        (setf AbsDepth Absolute)
        ;; Relative settings are void
        (setf RelWidth 1)
        (setf RelDepth 1))
        ((eq PrimaryDimensionality 2)
        (setf AbsWidth Absolute)
        (setf RelWidth 1))
        (t (format t "Width change is undefined for objects that are 0 or 1D-%")))
        ;; relative size change
    (cond
        ((eq PrimaryDimensionality 3)
        (setf RelWidth (* RelWidth ChangeSize))
        (setf RelDepth (* RelDepth ChangeSize)))
        ((eq PrimaryDimensionality 2)
        (setf RelWidth (* RelWidth ChangeSize)))
        (t (format t "Width change is undefined for objects that are 0 or 1D-%")))))))
```

FIG. 7.5. "Width" scaling is sensitive to dimensionality of axis. Objects with 3-D primary dimensionality are scaled in both the width and depth. Objects with 2-D primary dimensionality, assuming the generating axis is one of the two, are scaled in the width only. Used in definition of *slim, fat,* and so forth.

ratio within a range. "A 1-inch-wide oval acorn" is also 1 inch wide with length varying to maintain an aspect ratio to satisfy the constraints of being oval. "A 1-inch-wide oval leaf" does not change its depth by virtue of the 2-D primary dimensionality.

3.4. Shape Analogy

Shapes are often described by analogy to other objects. These analogies allow a speaker to convey information about many parameters of shape with minimal linguistic effort. In such cases the analogy replaces the prototype as the point of departure for difference description. A leaf can be "tear shaped," "mitten shaped," "spearhead shaped," and so on. Linguistic markers are used to describe what aspects of the object of the analogy are important. In this case, the term *shaped* is added. The linguistic structures of these analogies include the figure, ground, and a dimension marker. The figure is the focus and is expressed as the syntactic subject. The figure is the component with the shape that is unknown to the listener. The ground is the component whose shape is being inherited by the figure.

In VerbalImage, the phrase "X-shaped" means that the figure inherits the geon shape description and associated mesh from the ground but not the metric values. Cross section and taper are inherited but not the length, width, and height. The ratio of these metrics, however, will be preserved. The length along the main axis of the figure will not be altered by the analogy, but the width and height will be reset to reflect the ground ratios. Any object in the lexicon with a geon shape may be used in a shape analogy.

The linguistic analogy constructs of "looks like" or "resembles" will be processed in VerbalImage. In these analogies, one object is said to look like another. Usually the analogy is qualified along some dimension. For example, in *Trees of North America* (Brockman, 1968) the following description of Black Oak can be found: "Black Oak *(Quercus velutina)* leaves resemble those of Northern Red Oak but have 5 to 7 lobes separated by variable sinuses and are coppery with axillary tufts of hair below" (p. 126).

The shape description for the Black Oak leaf is the same as for the Northern Red Oak except for the modifications specified. From the example, it is easy to see that the analogy applies to the current level of description and to the parts of the focus object, in this case the lobes and sinuses.

VerbalImage is justifiably limited in that it does not possess general world knowledge about the shape of many common world objects.

3.5. Image Generation

G transforms symbolic physical objects, O_C, to graphic form, O_G. In VerbalImage G transforms the parameterized representation of objects into the object-oriented graphic language of YART, $G(O_C) \Rightarrow O_G$. YART[6] (Beier, 1994a, 1994b) is public domain, extensible, object-oriented C++ kernel with consistent interpretative language binding (Tcl) and supports ray tracing, radiosity, and shading.

There is a one-to-one correspondence between O_C and O_G at the object level but not at the attribute level. G maps each object in the domain model to a YART object. In this implementation the natural language processing subsystem resides on one machine. YART code is passed to a second architecturally appropriate machine.

The form of the symbolic domain model, D, and the resulting instance O_C is independent of the display representation. In this version of VerbalImage all objects are represented as polyhedrons. The only requirements for the graphic representation are that the functions G and I are defined under the representation. For example there must be an I function Length() that can return the length of the graphic object O_G independent of its representation as a polyhedron, analytic function, NURB, or other form.

The attribute list, A_I, generated by I serves as the index vector for image database retrieval. These attributes are quantitative versions of Biederman's (1987) attributes of cross-sectional edge curvature, symmetry, size, axis curvature, length, and aspect ratio. For particular object classes, some of these attributes, those that have little or no discriminate value, are excluded from the index vector for the object class. For example, all acorns have curved cross-sectional edges with rotational and reflectional symmetry of the cross section, so these terms are dropped from the index. In the VerbalImage architecture, verbal to image mapping, $S(U,D) \Rightarrow O_C$, is independent from symbolic to image mapping, $G(O_C) \Rightarrow O_G$ and independent of image to index mapping, $I(O_G) \Rightarrow A_I$.

4. CONCLUSIONS

VerbalImage demonstrates the integration of representations and procedures that are drawn from the psychology of visual perception and cognitive linguistics. These theories are extended to encompass a subcategorization task of object recognition from verbal description. Visual prototypes and exemplars prove to be useful constructs for the understanding of the verbal description of objects. These prototypes and exemplars are parameterized to form symbolic abstractions that are crucial for interpretation of language. These parameters include the primary and secondary dimensionality of the objects as well as the directedness of the three main

---

[6]For source and documentation for YART see http://metallica.prakinf.tuilmenau.de/ GOOD.html. Necessary extensions may be obtained from the author of this chapter.

axes of the object. The parameters that prove useful for language understanding are the same as those that are useful for image indexing. The parameterization for image indexing and retrieval is different from that for language understanding. The indexing parameters need to encode finer distinctions that are derived from the prototypes of the domain model rather than information explicitly provided in language.

Much work must be done to expand linguistic coverage and the size of the very modest database being indexed here.

ACKNOWLEDGMENTS

This research was sponsored, in part, by a U.S. Department of Education, Title IIb Award #R036B30088-94, awarded to the University of Pittsburgh, Project Director Steve Hirtle. The contents of this article were developed under a grant from the Department of Education. However, those contents do not necessarily represent the policy of the Department of Education and you should not assume endorsement by the federal government.

REFERENCES

Beier, E. (1994a). Object-oriented design of graphical attributes. *Proceedings of the 4th EuroGraphics Workshop on Object-Oriented Graphics* (pp. 41–50). Sintra, Portugal.

Beier, E. (1994b). *Objektorientierte 3D-Grafik* [Object-oriented 3D graphics]. Bonn: International Thomson Publishing.

Biederman, I. (1987). Recognition by components: A theory of human image understanding. *Psychological Review, 94*(2), 115–147.

Brockman, C. F. (1968). *Trees of North America*. New York: Golden Press.

Herskovits, A. (1985). Semantics and pragmatics of locative expressions. *Cognitive Science, 9*(3), 341–378.

Herskovits, A. (1986). *Language and spatial cognition: An interdisciplinary study of the preposition in English.* New York: Cambridge University Press.

Hoffman, D. D., & Richards, W. (1984). Parts of recognition. *Cognition, 18,* 65–96.

Jackendoff, R. S. (1991). Parts and boundaries. *Cognition, 41,* 9–44.

Landau, B., & Jackendoff, R. (1993). "What" and "where" in spatial language and spatial cognition. *Behavioral and Brain Sciences, 16,* 217–265.

Marr, D. (1982). *Vision: A computational investigation into the human representation and processing of visual information.* San Francisco: Freeman.

Ó Nualláin, S., Farley, B., & Smith, A. G. (1994). The Spoken Image System: On the visual interpretation of verbal scene descriptions. In P. McKevitt (Ed.), *Proceedings of the Workshop on Integration of Natural Language and Vision Processing* (pp. 36–39).

Rosch, E., & Mervis, C. (1975). Family resemblances: Studies in the internal structure of categories. *Cognitive Psychology, 7,* 573–605.

Tversky, B., & Hemenway, K. (1984). Objects, parts, and categories. *Journal of Experimental Psychology, 113*(2), 169–191.

8

Lexical Allocation in Interlingua-Based Machine Translation of Spatial Expressions

Clare R. Voss
Bonnie J. Dorr
M. Ülkü Şencan
University of Maryland

Given a spatial expression, or its computational semantic form, how is the expression's spatial semantics to be allocated *lexically*, that is, among the expression's entries in the lexicon? In interlingua-based machine translation (MT) research, *lexical allocation* is the problem of *allocating* or subdividing a linguistic expression's full interlingual (IL) structure into the substructures that are *lexical* IL forms, that is, in the lexicon. Here we present our work developing IL forms and an *IL lexicon* for translating English spatial expressions into Turkish. We describe different ways in which spatial information is allocated to lexical IL forms in our MT system, preruntime during lexicon construction and at runtime during lexical selection.

1. INTRODUCTION

In this chapter we report on our current research developing computational forms for the interlingua-based machine translation of spatial expressions. We frame this research in terms of the following problem of *lexical allocation* in natural language processing (NLP): Given a spatial expression, or its computational semantic form, how is its spatial semantics to be allocated among the expression's entries in the lexicon? For the particular NLP application of IL-based MT, the allocation problem refers to subdividing a linguistic expression's full interlingual structure into substructures, each of which corresponds to a lexicon entry's IL form, that is, a *lexical* IL form. The problem appears at two phases:

1. During construction of MT lexicon entries (before MT runtime): Given a spatial expression, how should the expression's spatial semantics be represented in an IL form and then be subdivided among the source language (SL) lexical elements present in the expression?

2. During lexical selection (at MT runtime): Given an IL form derived from a spatial expression, how should that IL form be subdivided among the target language (TL) lexical entries available in the MT system?

In this chapter we focus on the lexical allocation of spatial directional information in translations from English to Turkish. Turkish has postpositions and case markings that differentiate simple directions from goal-directed paths, giving us insights into these components of spatial semantics. We examine the contrast among expressions whose verb phrases contain both a motion verb (spatial placement or displacement) and a directional adposition (particles in English and postpositions in Turkish),[1] while they display distinct surface co-occurrence patterns. We describe our work developing IL forms and an *IL lexicon* for these two types of lexical items and the spatial relations appearing in their co-occurrence patterns.

2. SPATIAL SEMANTICS IN IL FORMS

We use the following terms to distinguish among different ways in which spatial information appears in the IL form of a linguistic expression:[2]

- *Spatial object-functions* include the geometric description functions of Herskovits (1986) and the schematization of objects of Jackendoff (1991).[3] These may take an object and yield a *spatial entity* (e.g., place), or they may apply recursively to a spatial entity. For example, *the*

---

[1]Here *adpositions* include prepositions, postpositions, and particles.

[2]This set is not comprehensive with respect to natural language spatial semantics (Dorr & Voss, 1993). It is beyond the scope of our chapter to address the complexities of functional and pragmatic levels of representation. For an approach that addresses these levels, including the geometric one we examine here, see Aurnague and Vieu (1993).

[3]Herskovits identified several object idealizations, parts, forms, volumes, axes and projections as *geometric description* (GD) functions on an object O at a time i. Jackendoff's including and extracting functions are not all strictly spatial schematizations.

top of the hill is a spatial entity where the function *top of* has an object, *hill*, as its argument.[4]

- *Spatial predicates* capture the positions of two entities relative to each other.[5] That configuration of positions in the real, three-dimensional (3-D) physical world may be assessed at a moment in time or over some duration, depending on the spatio-temporal properties of each entity. For example, in the case of *they danced in D.C.*, the spatial predicate for *in* locates the event entity for *they danced* relative to the place entity for *D.C.*[6]

- *Spatial situations* are events and states in the real, 3-D physical world, involving the motion, position, or spatial configuration of their participants.[7] For example, the verb in *he shelved the book* conveys a stereotyped motion event in addition to an incorporated object for *shelf*.

Note that more than one of these types of spatial IL forms may be allocated to the same lexical name. Consider, for example, the direction *up*, a lexicalization of a vector, and words sharing that semantics with a prefix *up-*. As a directed one-dimensional axis with an origin, a vector is clearly a geometric abstraction. However, vectors also appear in the semantics of ordinary, natural language as spatial object-functions and predicates. When we say that something is located *upstate* or *uphill*, then it is in *the upper part of* the state or toward or at *the top of* the hill. These *up*'s contain spatial object-functions on an entity (here, *state* or *hill*), yielding the uppermost portion of the spatial entity, a place. On the other hand, when we say that a canoe is headed *upriver* or a bird is headed *upwind*, then we know the direction of the canoe and bird relative to other entities, the river and the wind, respectively. These *up*'s contain spatial predicates on those entities, yielding a directed path.

---

[4]Reyero-Sans and Tsujii (1994) have a similar type they identified as *structural terms*, in contrast to *relational terms*. Aurnague and Vieu (1993) used the term *internal localization nouns* for lexical elements that specify different portions of an entity.

[5]We use *predicate* here in the logico-mathematical sense of the term, not its linguistic usage of a verb phrase consisting of a verb and only its internal argument(s). Unless otherwise noted, we use *predicate* for the full structure of a predicator and its arguments.

[6]Most typical examples of spatial predicates in the literature are prepositions (e.g., see Retz-Schmidt, 1988, for an excellent survey). For a presentation that goes beyond the syntactic category of prepositions, see Heine, Claudi, and Huennemeyer (1991).

[7]There may also exist another category distinct from, or a subset of, spatial situations. It carries spatial substructure and semantic aspect, but not temporal location, which it derives from another phrase. For example, in *he came into the room, running*, the action of *running* is contemporaneous with the *come* motion, deriving its temporal location (past) and directionality (*into the room*) from the main phrase in the expression.

Given this framework, we now look at spatial expressions and examine how these types of IL forms are needed to capture the spatial semantics in translating from English to Turkish.

3. DATA SET

In this section we present the co-occurrence patterns of spatial motion verbs and directional adpositions together in verb phrases. We first look at the patterns in English expressions and then at translations of these expressions into Turkish. Our goal is to identify expressions close in overall sentential meaning but whose individual verbs vary in their co-occurrence patterns with adpositions. Accounting for this contrast—close sentential semantics but distinct lexical co-occurrence patterns—requires allocating distinct lexical semantics to the verbs.

3.1. Patterns in English

In the expressions that follow, '*' and '*/?' mark sentences judged un-grammatical and questionable (i.e., an unsure judgment) by native English speakers, respectively. '(uM)' identifies an implicit upward motion (place-ment or displacement) in a grammatical sentence.

Yesterday we watched as the crane operator . . . . . .

(1) (i)   (uM) elevated the new railroad tracks.
    (ii) *elevated **up** the new railroad tracks.
    (iii) *elevated **down** the new railroad tracks.

(2) (i)   (uM) lifted the new railroad tracks.
    (ii) (uM) lifted **up** the new railroad tracks.
    (iii) */?lifted **down** the new railroad tracks.

(3) (i)   *put the new railroad tracks.
    (ii) put **up** the new railroad tracks.
    (iii) put **down** the new railroad tracks.

(4) (i)   moved the new railroad tracks.
    (ii) moved **up** the new railroad tracks.
    (iii) moved **down** the new railroad tracks.

Table 8.1 summarizes the co-occurrence patterns from these sentences. The '*' and '?' are used as in the preceding expressions. For example, the combination of *elevate* and *up* is ungrammatical. We introduce here a '+'

TABLE 8.1
English Co-Occurrence Classes

Obligatorily Lexically Implicit	Lexically Implicit and Optionally Explicit
Verbs have an inherent spatial direction that cannot be lexicalized explicitly e.g., *elevate* [* UP] [* DOWN]	Verbs have an inherent default direction that can also be explicit. e.g., *lift* [(+) UP] [*/? DOWN]

Obligatorily Lexically Explicit	Optionally Lexically Explicit
Verbs have no inherent direction, yet require an explicit location as an argument e.g., *put* [+ UP] [+ DOWN]	Verbs have no inherent specific direction, only semantics of spatial motion e.g., *move* [(+) UP] [(+) DOWN]

and '(+)' for grammatical combinations where the argument (not the specific direction) is obligatory and optional, respectively.

3.2. Patterns in Turkish

Here we present Turkish translations of the English sentences from the previous subsection and then summarize the co-occurrence classes for the verbs involved. In several of the following sentences, the English word *up* may translate into the Turkish root form *yukarı* as either a simple vector *upward* or as a goal-marked path *to a place that is up/higher*. In the former case of the vector direction *upward*, the word *yukarı* is used. We refer back to this sense as *y-1*. In the latter case of a path to or toward some goal,[8] Turkish has two forms: (a) *yukarı* or (b) *yukarıya*. We refer back to these related senses as *y-2* and *y-3*.[9]

    E: Yesterday we watched as the crane operator....[10]

    T: Dün, vinç operatörünün . . . seyrettik.

         "Yesterday, crane operator . . . watched."

   (5) (i) E: **elevated** the new railroad tracks.

           T: yeni demiryolu raylarını **yükseltişini**

              "new railroad tracks elevated"

---

[8]In general the *to* meaning is stronger than the *toward*.

[9]This distinction is slippery indeed, if not elusive, for non-Turkish speakers. The *-a* suffix on *yukarıya* marks it as a noun. When *yukarı* is read with sense *y-2*, it has no overt marker, only an implicit goal.

[10]We use E and T to designate the original English sentence and the Turkish translation of E. An extra line with the word-for-word English translation of T preserving Turkish word order appears in quote marks as needed.

(ii) E: elevated *up/*down the new railroad tracks.
    T: yeni demiryolu raylarını *yukarı/*aşağı yükseltişini[11]

(6) (i) E: **lifted** the new railroad tracks.
      T: yeni demiryolu raylarını **kaldırışını**

  (ii) E: lifted **up**/?*down the new railroad tracks.
      T: yeni demiryolu raylarını (yukarı/yukarıya) / *(aşağı/aşağıya) kaldırışını

(7) (i) E: *put the new railroad tracks.
      T: yeni demiryolu raylarını **koyuşunu**
         "new railroad tracks **placed**"

  (ii) E: put up/down the new railroad tracks.
      T: yeni demiryolu raylarını (yukarı/yukarıya) / (aşağı/aşağıya) koyuşunu
         "new railroad tracks up/down placed"

(8) (i) E: **moved** the new railroad tracks.
      T: yeni demiryolu raylarını **hareket ettirişini**
         "new railroad tracks **motion do-caused**"[12]

  (ii) E: moved **up/down** the new railroad tracks.
      T: yeni demiryolu raylarını (yukarı/yukarıya (doğru)) / (aşağı/aşağıya (doğru)) hareket ettirişini[13]

In sentences (6ii) and (8ii), the translation of *up* to *yukarı* is ambiguous, meaning either *y-1* or *y-2*, as described earlier. However, in sentence (7ii), only the meaning *y-2* is acceptable; the meaning *y-1* is strictly ruled out. That is, the *koy* motion (in the closest translation for the English *put*) must be goal directed.

In Table 8.2, the same marking conventions are used as in section 3.1. We see here that for each Turkish co-occurrence pattern, there is an equivalent one in section 3.1 for English (though not vice versa).

3.3. Lexical Allocation in the Data Set

We have limited our presentation here to these verb-adposition distribution classes, although there are others.[14] Our aim has not been to provide a complete enumeration of verb classes and verb-adposition alternations.[15] Rather, the

---

[11]Translation of this ungrammatical sentence was based on sentence 5(i) and inserting the spatial postposition.

[12]This causative sense of *move* is distinct from the reflexive (self-caused) sense.

[13]Adding *doğru* forces a goal-directed path reading, suggesting that a *y-2* reading of *yukarı* is possible.

[14]Consider the verb *pick*. We can *pick up* a box but not *pick down* a box. Furthermore, if we merely *pick* a box, the verb is possessional rather than spatial.

TABLE 8.2
Turkish Co-Occurrence Classes

Obligatorily Lexically Implicit	Lexically Implicit and Optionally Explicit
T: *yükselt* [* YUKARI] [* AŞAĞI] 'elevate' [* UP] [* DOWN]	T: *kaldır* [(+) YUKARI] [* AŞAĞI] 'lift' [(+) UP] [* DOWN]
Obligatorily Lexically Explicit	Optionally Lexically Explicit
T: closest translation of *put* is *koy* (see Optionally Lexically Explicit)	T: *koy* [(+) YUKARI] [(+) AŞAĞI] 'place' [(+) UP] [(+) DOWN] T: *hareket ettir* [(+) YUKARI] [(+) AŞAĞI] 'move' [(+) UP] [(+) DOWN]

goal has been to identify expressions close in overall sentential meaning but whose individual verbs vary in their co-occurrence patterns with the same adpositions. Accounting for this contrast—co-occurrence patterns of motion verbs with the same directional adposition—provides a natural test for different solutions to the lexical allocation problem, that is, subdividing sentential meaning in distinct ways for each verb-adposition pair.

4. ALLOCATION DURING MT LEXICON CONSTRUCTION

This section examines the problem of lexical allocation during MT lexicon construction, before MT runtime.[16] We use *extralexical operations*[17] to capture the compositional properties of lexical elements in the sentences in section 3. After sketching out these operations, we describe our basic lexical IL forms and the annotations on these forms for computing the operations. We end this section with a discussion of additions needed to handle the Turkish data with our IL forms.

4.1. Extralexical Operations

The extralexical operations we describe here serve as a bridge from the linguistic data of co-occurrence classes in section 3 to the specific com-

---

[15]Levin (1993) provided such an extensive list for English, for example, and Talmy (1985) examined related issues cross-linguistically.

[16]MT lexicon entries can also be created at runtime, for example, Onyshkevych and Nirenburg (1995).

[17]We define an IL's syntax in terms of (a) lexical IL forms within language-specific lexicons and (b) algorithms for creating and decomposing the instantiated *pivot* representation (Voss & Dorr, 1995). The *extralexical operations* are an abbreviated version of (b).

putational approach—annotations on IL forms—spelled out further in section 4.2. With respect to the verbs in the last section, we say that the IL form for the verb and the IL form for *up* are related as follows:

- Obligatorily Lexically Implicit
  e.g.: an IL form for *up* is internal to the IL form for *elevate*,
  elevate's IL form **blocks** attachment with another *up*'s IL form
- Lexically Implicit and Optionally Explicit
  e.g.: an IL form for *up* is internal to the IL form for *lift*,
  lift's IL form may **overlap** with another *up*'s IL form
- Obligatorily Lexically Explicit
  e.g.: no IL form for *up* is internal to the IL form for *put*,
  put's IL form must be **filled** in its locational argument position, such as with *up*'s IL form
- Optionally Lexically Explicit
  e.g.: no IL form for *up* is internal to the IL form for *move*,
  move's IL form may be **filled** in its spatial argument position, such as with *up*'s IL form

More generally, given an IL form for a spatial expression, we specify that the IL form allocated for the verb, call it V, and the IL form allocated for *up*, call it P, may be related in one of the following ways: (a) **block:** V blocks P from attaching to any node in V, (b) **overlap:** V permits P to attach and share subtree rooted at a nonleaf node in V, and (c) **fill:** V permits P to attach at a leaf node in V.

4.2. Computational Forms

Our approach in developing IL forms derives from the lexical conceptual structures (LCSs) of Jackendoff (1983, 1990, 1991).[18] The LCS framework consists of three independent subsystems: fields, conceptual constituents, and boundedness/aggregation properties. Only the first two are currently a part of our IL forms, as shown in Fig. 8.1. The LCS fields (spatial, temporal, possessional, identificational, and others) are motivated by well-known observations of *lexical parallelism*, where the same lexical item has parallel or related meanings in other semantic fields.

The conceptual constituents in the second LCS subsystem are variants on predicate-argument structures. They are typed by one of a small set of ontological categories (Thing, State, Event, Place, Path, Property, and Amount) and the internal structure of each constituent may decompose

---

[18]See Dorr (1993) for details of a MT system whose IL is LCS-derived. Recently the LCS framework has been used by others for French, for example, Pugeault, Saint-Dizier, and Monteil (1994) and Verrière (1994).

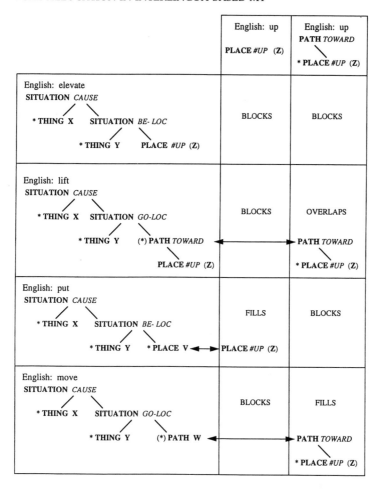

FIG. 8.1. IL forms for English verbs and *up*.

into other conceptual constituents. The predicate primitives are sub-scripted by field (such as LOC for the spatial field) in addition to being typed by category. Although all primitives may not appear in all fields (for a particular language), Jackendoff did make the claim that the constituent structures do generalize across fields. In particular, he adapted a *localist* view, claiming that the formalism for encoding constituents in the spatial field, at some level of abstraction, generalizes to other fields.[19] At the end of this chapter we present a few examples that support this approach—we show that the distinction we build into our IL forms for the spatial field to handle the Turkish data extends to other fields as well.

---

[19]A localist, or localist-related, approach is by no means unique to Jackendoff. See, for example, among many others, Anderson (1971), Heine et al. (1991), and Langacker (1987).

TABLE 8.3
Annotations on Lexical IL Forms By Co-Occurrence Class

Co-Occurrence Class	Extralexical Operation	Marker	Attachment Site
Obligatorily Lex. Implicit	block	none	not applicable
Lex. Implicit & Optionally Explicit	overlap	(*)	internal
Obligatorily Lex. Explicit	fill	*	external
Optionally Lex. Explicit	fill	(*)	external

In the lexical IL forms of Fig. 8.1, two types of markers or annotations appear on leaf and nonleaf nodes in order to implement the extralexical operations sketched earlier. The '*' marker at a node indicates that an obligatory attachment by another IL form occurs at that site. The '(*)' marker indicates an optional IL form attachment may occur at that site. The relation between the co-occurrence classes, the extralexical operations, and the markings on the IL forms appears in Table 8.3. The annotations on lexical IL forms are used at MT runtime during the analysis and generation phase to guide, respectively, the composition and decomposition of the full IL form corresponding to the input spatial expression. The annotations are language-specific and are removed from the fully composed IL form when the analysis phase is complete.[20]

4.3. Lexical Construction for Turkish Data

In the summary tables of section 3, we saw that for three of the four co-occurrence classes, the Turkish verb used to translate the English verb had the same distribution pattern as its English counterpart. The one case where this did not occur required further examination. As noted in that section, the Turkish verb *koy* that comes the closest to English word *put* cannot take as its argument the *y-1*, or simple vector sense of *up*. Although *koy* readily takes a Place type argument and will not take a *y-1* Path type argument, we discovered that it would accept *y-2* and *y-3*, the goal-marked Path senses, as arguments.

This was a surprising fact: Why would one Path be acceptable to a verb that generally takes a Place argument whereas another Path was not? It turned out the goal-marked Path sense was acceptable only under the coerced interpretation that these were Places.[21] That is, for the Turkish

---

[20]The details of these algorithms are discussed in Dorr (1993).

[21]For comparison, consider the English sentence *he lives through the tunnel*. This means he lives somewhere that is through, that is, beyond the tunnel. Here the verb *live*, which requires a Place as its argument type, coerces the Path *through the tunnel* into a Place by treating it as though embedded within an indefinite relative clause *somewhere that is through the tunnel*.

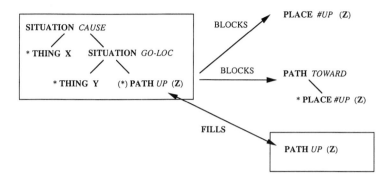

FIG. 8.2. Additional IL forms.

equivalent of *put*, simple vector Paths are apparently not coercible into Places, whereas goal-marked Paths are.

In order to capture this contrast, we have distinct IL forms for simple vectors and goal-marked paths. We hypothesize that *koy*, in looking for a Place as its argument, will accept and coerce a goal-marked Path into a Place, but not a simple vector, precisely because the former contains a Place whereas the latter does not. In other words, coercion to a Place is an operation that selects for boundable entities and thus rejects simple vectors.

The lexical IL forms to handle these data appear in boxes in Fig. 8.2. In particular, a new Path predicate *UP* was introduced for the simple vector sense of *up*. Previously, consistent with Jackendoff's limited set of five Path primitives, the MT lexicon held one entry for the English *up*, a Path predicate *TOWARD* with an argument that was a Place object-function *#UP*. As a side effect of this change, the IL syntax now also allows for new spatial situations containing this new Path predicate for simple vectors.

5. ALLOCATION DURING LEXICAL SELECTION

The *allocation* analysis described earlier occurs at MT lexicon definition-time, when the IL developer decides what IL forms to allocate to the lexical entries.[22] Here we describe our work building *IL lexicons* for lexical selection at MT runtime. In an IL-based MT system, the process of lexical selection is one part of the generation phase that follows the construction of a *pivot* IL form from the input source language expression. The pivot IL form must be subdivided into IL forms that correspond to lexical IL forms in the target

---

[22]The allocation operations execute at runtime, although they originate in the IL forms created at definition-time.

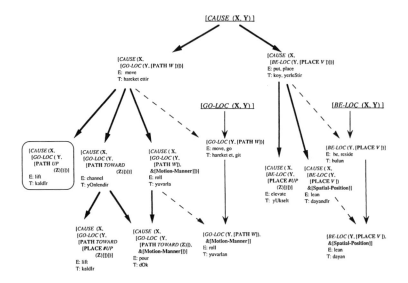

FIG. 8.3. Subgraph of *IL Lexicon* for spatial situations.

language's lexicon. We split the selection process into several components. Of interest here is the one that, given a subpart of the pivot IL form, will find a range of TL lexical items whose IL forms cover or approximate that given subpart. Following DiMarco, Hirst, and Stede (1993), we call this component a *lexical option finder*. Our approach has been to build an *IL lexicon* for the finder, a reverse index into the TL lexicons.

5.1. IL Lexicons

Our IL lexicon is a hierarchical data structure that organizes the space of IL forms extracted from the MT system's lexicons.[23] Its purpose is to structure the search space of lexical IL forms for the lexical option finder by providing a reverse-index into language-specific MT lexicons. We note here in passing that the IL lexicon also *grounds* the IL forms by coindexing the IL primitives in the KR system. Figure 8.3 shows a subgraph of the the spatial situations in the IL lexicon and Fig. 8.4 shows a subgraph of the spatial predicates and object-functions in the IL lexicon.

The nodes of an IL lexicon are IL forms corresponding to at least one lexical item's entry in one of the MT lexicons. The nodes are connected by links based on the structural properties of the lexical IL form identi-

---

[23]We use the phrase *MT lexicons* to refer to the natural language-specific lexicons in MT systems that are organized by language-specific entries. In an earlier description of this work, we identified the IL lexicon as a *concept-based lexicon* (Dorr, Voss, Peterson, & Kiker, 1994).

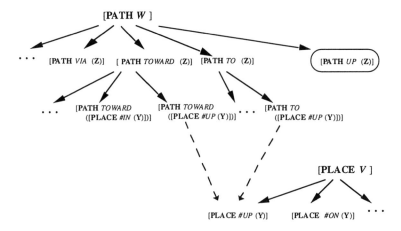

FIG. 8.4. Subgraph of *IL Lexicon* for spatial predicates and object-functions.

fying that node. In Figs. 8.3 and 8.4, the bold arrows stand for structural subsumption relations between node identifiers. The dashed arrows are *reduction links* connecting one lexical IL form to another that matches one of its substructures. Attached to each node are reverse-index pointers to its lexicalizations, that is, the lexical entries in the MT lexicons in the system. In other words, the nodes are not lexical items per se, but rather concepts defined only in terms of an IL form identifier without language-specific annotations (such as the *-markers described earlier).[24]

The lexical option finder traverses the IL lexicon following these links, using nodes with pointers to TL lexicalizations to generate the range of lexical items for other components in the lexical selection process.[25] Thus far in our IL lexicon the nodes have been lexical IL forms with a function- or predicate-argument structure. We have relied on the ontology in a knowledge representation system for structuring relations among objects identified by simple IL forms.

5.2. Lexical Selection With Turkish Data

In the IL lexicon Figs. 8.3 and 8.4, the IL forms that were added in to account for the distinction between simple vectors and goal-marked paths in Turkish are circled. The Turkish simple vector for *up* labeled earlier as *y-1* is a lexicalization of the predicate [**PATH** *UP*(Z)]. The other Turkish

---

[24]Although this approach resembles that taken by Wu and Palmer (1994), who addressed related issues in lexical selection, our IL lexicon is structured by the syntax of our IL whereas their hierarchy is defined in terms of concepts that encode a multipart meaning representation.

[25]Further details are presented in Dorr et al. (1994).

senses of *up*, *y-2* and *y-3*, are lexicalizations of paths *TOWARD* and *TO* with the object-function Place argument *#UP*.

6. CONCLUSION AND FUTURE WORK

This chapter addresses the general problem of lexical allocation in IL-based MT in conjunction with the specific question of how to represent spatial directions in LCS-derived IL forms. With respect to MT research, we show that, in our IL-based system, the two operationally distinct phases of lexical construction (preruntime) and lexical selection (runtime) are implicitly interdependent. Within our framework, these both are problems in lexical allocation and depend on the syntax of the IL. In particular, the IL syntax formalism defines (i) the structures for the spatial information to be allocated to lexical items and (ii) the search space of the IL lexicon to be traversed during lexical selection. Our next step is to build support tools to specify a lexicalized IL grammar and guide the building of lexical IL forms and an IL lexicon.

With respect to spatial expressions, we have pursued the hypothesis that directions lexicalize in two ways, using spatial object-functions and spatial predicates. We use these to distinguish (i) goal-marked paths from (ii) simple vectors in our lexical IL forms, a contrast needed for translation into Turkish.[26] In future work we will expand our data set in order to test this hypothesis further. We note in concluding that our work remains consistent within the localist framework: Both of our Path interpretations of the English word *up* extend to nonspatial fields, as shown in the following sample sentences where the *up* semantics may be lexically explicit in the (a) sentences or implicit in the (b) sentences.

(i) temporal: (a) They moved up the deadline.
                            (b) They advanced/delayed the deadline.
     identificational: (a) Her temperature went up (and stayed there).
                            (b) Her temperature peaked.
(ii) temporal: (a) He sped up the car.
                            (b) He accelerated the car.
     identificational: (a) The rocket went up (and up).
                            (b) The rocket rose/soared.

ACKNOWLEDGMENTS

This research was supported, in part, by the Army Research Office under contract DAAL03-91-C-0034 through Battelle Corporation, by the National

---

[26]The two forms are: (i) **[PATH** *TOWARD* **[PLACE** *#UP*(x)]] with the spatial object-function *#UP*, and (ii) **[PATH** *UP*(z)] where the spatial predicate is *UP*.

Science Foundation under grants NYI IRI-9357731, NSF/CNRS INT-9314583, by the Alfred P. Sloan Research Fellow Award BR3336, and by the Army Research Institute under contract MDA-903-92-R-0035 through Microelectronics and Design, Inc.

REFERENCES

Anderson, J. (1971). *The grammar of case: Towards a localist theory.* Cambridge, England: Cambridge University Press.

Aurnague, M., & Vieu, L. (1993). A three-level approach to the semantics of space. In C. Zelinsky-Wibbelt (Ed.), *The semantics of prepositions: From mental processing to natural language processing* (pp. 393–439). Berlin: Mouton de Gruyter.

DiMarco, C., Hirst, G., & Stede, M. (1993). The semantic and stylistic differentiation of synonyms and near-synonyms. In *Working notes for the AAAI Spring Symposium on Building Lexicons for Machine Translation* (Tech. Rep. No. SS-93-02, pp. 114–121). Stanford, CA: Stanford University.

Dorr, B. (1993). *Machine translation: A view from the lexicon.* Cambridge, MA: MIT Press.

Dorr, B., & Voss, C. (1993). Machine translation of spatial expressions: Defining the relation between an interlingua and a knowledge representation system. In *Proceedings of the AAAI* (pp. 374–379). Menlo Park, CA: AAAI.

Dorr, B., Voss, C., Peterson, E., & Kiker, M. (1994). Concept based lexical selection. In *AAAI 1994 Fall Symposium on Knowledge Representation for Natural Language Processing in Implemented Systems* (pp. 21–30). Menlo Park, CA: AAAI.

Heine, B., Claudi, U., & Huennemeyer, F. (1991). *Grammaticalization: A conceptual framework.* Chicago: University of Chicago Press.

Herskovits, A. (1986). *Language and spatial cognition.* Cambridge, England: Cambridge University Press.

Jackendoff, R. (1983). *Language and cognition.* Cambridge, MA: MIT Press.

Jackendoff, R. (1990). *Semantic structures.* Cambridge, MA: MIT Press.

Jackendoff, R. (1991). Parts and boundaries. In B. Levin & S. Pinker (Eds.), *Lexical and conceptual semantics* (pp. 9–45). Cambridge, MA: Blackwell Publishers.

Langacker, R. (1987). *Foundations of cognitive grammar: Vol. 1. Theoretical prerequisites.* Stanford, CA: Stanford University Press.

Levin, B. (1993). *English verb classes and alternations: A preliminary investigation.* Chicago: University of Chicago Press.

Onyshkevych, B., & Nirenburg, S. (1995). A lexicon for Knowledge-Based MT. *Journal of Machine Translation, 10*(1–2), 5–57.

Pugeault, F., Saint-Dizier, P., & Monteil, M. G. (1994). Knowledge extraction from texts: A method for extracting predicate-argument structures from texts. In *Proceedings of Fifteenth International Conference on Computational Linguistics* (pp. 1039–1043). Morristown, NJ: Association for Computational Linguistics.

Retz-Schmidt, G. (1988). Various views on spatial prepositions. *AI Magazine, Summer,* 95–105.

Reyero-Sans, I., & Tsujii, J. (1994). A cognitive approach to an interlingua representation of spatial descriptions. In *AAAI Workshop on Integration of Natural Language and Vision Processing* (pp. 122–130). Menlo Park, CA: AAAI.

Talmy, L. (1985). Lexicalization patterns: Semantic structure in lexical forms. In T. Shopen (Ed.), *Language typology and syntactic description 3: Grammatical categories and the lexicon* (pp. 57–149). Cambridge, England: Cambridge University Press.

Verrière, G. (1994). Manuel d'utilisation de la structure lexicale conceptuelle (LCS) pour représenter des phrases en français [Users' manual for representing French sentences in Lexical Conceptual Structures (LCS)]. (Research Note, IRIT). Toulouse, France: Université Paul Sabatier.

Voss, C., & Dorr, B. (1995). Toward a lexicalized grammar for interlinguas. *Journal of Machine Translation, 10*(1–2), 143–184.

Wu, Z., & Palmer, M. (1994). Verb semantics and lexical selection. In *Proceedings of the 32nd Annual Meeting of the Association for Computational Linguistics* (pp. 133–138). Morristown, NJ: Association for Computational Linguistics.

9

Schematization

Annette Herskovits[1]
Boston University

Almost every study of spatial prepositions mentions *schematization* (though it may not use this exact term). Here is Talmy's (1983) characterization of schematization: "a process that involves the systematic selection of certain aspects of a referent scene to represent the whole, disregarding the remaining aspects" (p. 225). Here is a citation from my own work (Herskovits, 1986): "[T]here is a fundamental or canonical view of the world, which in everyday life is taken as the world as it is. But language does not directly reflect that view. Idealizations, approximations, conceptualizations, mediate between this canonical view and language" (p. 2). Systematic selection, idealization, approximation, and conceptualization are facets of schematization, a process that reduces a real physical scene, with all its richness of detail, to a very sparse and sketchy semantic content. For expressions such as "The village is on the road to London." this reduction is often said to involve applying some abstract spatial relation to simple geometric objects: points, lines, surfaces, or blobs.

Work in artificial intelligence sometimes mentions schematization, but I know of no computational model of the use of spatial expressions in which it plays a significant role. Yet, schematization cannot be overlooked in modeling human abilities; it is most certainly a key to understanding both the strengths and limitations of spatial language.

---

[1]Presently at the Institute of Cognitive Studies at the University of California, Berkeley.

Schematization involves three distinguishable processes: abstraction, idealization, and selection. Abstraction, of course, is an essential characteristic of *all* linguistic meaning. Every linguistic category abstracts from the distinguishing characteristics of its individual members. In saying "Joe is running." we abstract away from particular distinguishing characteristics of Joe's running—speed, style, location, goal, and so on. Similarly, in saying "There is a tree lying across the road." we abstract away from the position of the tree along the road, the angle between tree axis and road axis, the position of the ends of the tree with respect to the road's edges, the width of the road, whether the road is in an horizontal plane or inclines steeply, and so on.

The facet of schematization particular to spatial language is *geometric idealization*. Spatial expressions conjure up points, lines, ribbons, and so forth, but the scene described does not usually include them; we "idealize" features of the real scene so they match these simple geometric objects. Idealization goes beyond abstraction: It implies a *mismatch* between the real geometric features and the categories in which we fit them. Thus the top surface of the road in the preceding example is (arguably) idealized to a ribbon, although it could be quite bumpy and of varying width.

Selection involves using a part or aspect of an object to represent the whole object, as with "the cat under the table," where the top of the table stands for the whole table. Including selections stretches the ordinary meaning of schematization—yet selections do fit Talmy's (1983) definition, and they commonly contribute to producing the reduced object geometry relevant to spatial expressions.

Marr (1979) wrote that an information-processing task must be understood at two levels: "The first, which I call the information theory of an information processing task, is concerned with what is being computed and why; and the second level, that at which particular algorithms are designed, with how the computation is to be carried out" (p. 19). With regard to the task of producing spatial expressions, we really still do not know what is being computed and why. We need to address the following questions: What evidence do we have that schematization takes place? Can we predict, for every expression and context, which schematization applies to the objects? How does the schematic representation of an object depend on (a) the intrinsic geometry of the object, (b) the preposition, and (c) the context?

Once we are clear about the conditions and effects of linguistic schematization, we can turn our attention to the relation between schematization and *nonlinguistic* spatial cognition, and ask: Is schematization a process performed for the purpose of linguistic expression, or does language simply sample schematic spatial representations constructed for nonlinguistic purposes?

I present some evidence and some nonevidence for particular schema-tizations, and suggest how they relate to nonlinguistic spatial cognition. We will see that, contrary to a common belief, prepositions do not always use representations of objects as points, lines, planes, or blobs.

1. TREATING OBJECTS AS POINTS:
A FALLACIOUS ARGUMENT

I follow Talmy in calling the two objects involved in a spatial relation the Figure and the Ground:

the lamp over the table
Figure   Ground

Talmy (1983) wrote that, typically, the prepositions "treat the focal object [the Figure] as a point or related simple form" (p. 234). This is a frequently expressed intuition, but it is not clear what "treating an object as a point" means. One justification often given for this claim is that if a preposition puts no constraint on the geometry of one of the objects related, then that object is treated as a point. As most prepositions do not restrict Figure shape, it follows that the Figure must generally be treated as a point.

Let us examine each step of this argument. Most of the prepositions are polysemous, but they can be roughly classified according to whether all their senses, or at least their most salient senses, are stationary or dynamic. Thus Table 9.1 lists the prepositions[2] sorted into two columns labeled respectively *Primarily Location* and *Primarily Motion*.[3]

Some stationary senses of the prepositions do in fact put constraints on Figure shape, for instance:

- Some stationary senses of the motion prepositions require the Figure to be a line:
  The snake lay across/along the trail.
- For one sense of "over," the Figure must be a surfacy object:
  The tablecloth lay over the table.

---

[2]The list is close to exhaustive; excluded are some specialized prepositions (e.g., *aboard*). Combinations whose meaning can be obtained by regular rules of semantic composition (e.g., *up into*) are not included.

[3]This classification is offered for expository convenience; it is not a theoretical statement. In fact, some prepositions may not fit clearly in any one class, e.g., *over* has equally salient location and motion senses, see "the lamp over the table" and "She walked over the hill." For an extended discussion of the classification, see Herskovits (1997).

TABLE 9.1
The English Spatial Prepositions

Primarily Location	Primarily Motion
at/on/in	across
upon	along
against	to/from
inside/outside	around
within/without	away from
near/(far from)	toward
next	up/down to
beside	up/down
by	into/(out of)
between	onto/off
beyond	out
opposite	through
amid	via
among	about
throughout	ahead of
above/below	past
under/over	
beneath	
underneath	
alongside	
on top/bottom of	
on the top/bottom of	
behind	
in front/back of	
left/right of	
at/on/to the left/right/front/back of	
at/on/to the left/right side	
north/east/west/south of	
to the east/north/south/west of	
on the east/north/south/west side of	

- For one sense of "throughout," the Figure must be a composite aggregate:

  There were blackbirds throughout the tree.

But, other than these few instances, the stationary senses of the prepositions put no constraint on Figure shape. So is the Figure treated as a point in all but these few cases?

There are clear other instances where the Figure is *not* treated as a point. The Figure can be infinite or unbounded:

The land beyond the river is fertile.

He contemplated the firmament above him.

In these sentences, the Figure has no boundary except where it touches the Ground. The firmament extends infinitely upward; the land stops somewhere, but this outer boundary is not part of the conceptualization— it is outside the scope of the mental eye. It would be absurd to assume that infinite or unbounded objects are seen or treated as points: Only bounded objects can be idealized to points. But even a bounded Figure is not necessarily seen as a point: Idealization of the figure to a point is irrelevant to

The orange juice is in the bottle.

The sheets are on the bed.

The Atlantic is between Europe and America.

So where is the difficulty? It is with the assumption that if a preposition places no constraints on the shape of one of the objects related, then that object is idealized to a point. Indeed, closer examination reveals that the premise does not entail the conclusion. The error may stem from the logical misstep: "If the prepositional predicate applies to objects of any shape, then its truth in particular cases can be assessed without referring to the object's shape." But that is clearly false.

Consider the preposition *in*. It does not restrict the Figure shape in any fashion; an object of any shape or even any dimensionality will do. In fact, the *selection restrictions* for the Ground are equally loose; any Ground shape will do, except a point—nothing can be *in* a point. But it certainly does not follow that Figure and Ground are treated or seen as points in uses of *in*. Ullman (1985) described various algorithms for deciding whether a point is *in* a closed curve. These, as one would expect, require full knowledge of the shape of the curve; hence the Ground (the curve) is not seen as a point. If the Figure is *extended*, we must also know the position of every one of its points to decide whether it is *in* the curve. Therefore we treat neither Figure nor Ground as points.

Perhaps objects are seen as "blobs" (which may be what Talmy [1983] meant by "related simple form")? A blob must be how we apprehend an extended object whose shape and extent are not known. Its representation could consist of the position of a center point of the object, together with the assumption that the object extends outward from this center to an undeterminate boundary. But without the precise extent of the Ground, we cannot *in general* decide whether an object is *in* it; in fact, we need also to know the exact region of space occupied by the Figure. So blobs won't do, nor will lines or planes.

In conclusion, the belief that *all* prepositions treat Figure and Ground as points, lines, planes, or blobs is unfounded; some do, and some do not. In fact, as we see later, some do in some uses and not in others. The precise regions occupied by Figure and Ground must sometimes be known to decide whether a preposition applies.

So, one main argument that prepositions idealize objects to points is untenable. Is there any reliable evidence of idealizations to a point?

2. FIGURE AS A POINT:
THE MOTION PREPOSITIONS

In the second column of Table 9.1 are prepositions whose primary senses serve to describe motion. In

The ball rolled across the street.
Figure   Ground

the Figure is the moving object, and *across* relates the trajectory of the Figure to a reference object or Ground, the street.

What sort of constraints do these prepositions put on the trajectory of a Figure? Each preposition defines, for any given Ground, a field of directed lines. In saying "The ball rolled across the street," we are asserting that the ball follows one of the lines of the field defined by "across the street." In Fig. 9.1 are diagrammatic representations of the field of directed lines for "across the road," "along the path," "around the park," and "toward the tower."

So the primary senses of the motion prepositions are all cast in terms of linear paths. A linear path is traced out by a point, or by a linear, deformable object sliding along its own axis, for instance (ignoring their cross section) a snake or a train. But such objects are rare, so we must somehow make use of predicates defined in terms of motion of a point to talk about motion of any extended object.

In fact, we generally ignore the complex geometry of the path described by a three-dimensional object moving though space. The path of a rigid object undergoing translation has the shape of a generalized cylinder, a kind of snake, but with possible overlaps, and not necessarily a circular cross section. If the object rotates on itself, changing its orientation as it translates, the volume described is typically too complicated to visualize precisely. But kinematics shows that the motion can be analyzed as a succession of infinitesimal motions, each combining a translation and a rotation around some axis.

However, for the most part, we lack the ability to represent subcategories of such paths involving different combinations of rotation and translation. Fine conceptual distinctions among the possible paths of a nonrigid object

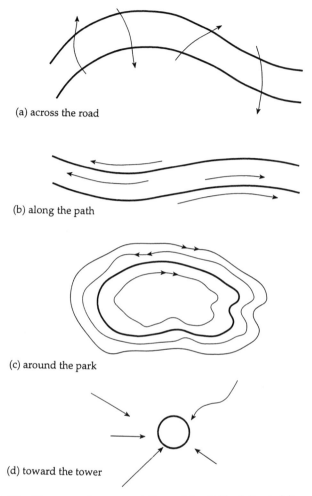

(a) across the road

(b) along the path

(c) around the park

(d) toward the tower

FIG. 9.1. Diagrammatic representations of the field of directed lines for salient motion senses of the prepositions *across*, *along*, *around*, and *toward*.

are even harder to make. The restriction is not in our awareness of the object's changing appearance as we watch it move; it is in our ability to analyze the motion, albeit unconsciously and nonverbally, that is, to note separable aspects of it. This ability is needed to form different subcategories of motion, and to assign a particular instance to a subcategory.

Because of this limitation, we typically idealize the motion of an object to that of a center point—probably its "centroid," or center of gravity assuming uniform density—ignoring rotations around the centroid. Note, for instance, that there exist no two prepositions which contrast only in that one entails a pure translation, and the other a translation accompanied by a rotation. A few verbs do describe trajectories involving different combinations of these motions—for example, roll, flip, tumble, turn, twirl, revolve—but the list is very short, which supports the claim that our ability to conceptually distinguish motions along these dimensions is limited. Consider "The child danced around the May Pole." Although we would typically assume that the child rotates on herself, the preposition does not require it. The child would still be "dancing around the May Pole" if the pole were on a stage and she continued facing the audience while dancing around it.

In a few expressions, the preposition does not describe the trajectory of the centroid. So

He rolled the log over.

The earth turns around its axis.

use prepositions specifying the motion of a point in a different way. Brugman (1988) provided a nice explanation for *over* in the first sentence. As the log rolls along the ground a distance equal to half its circumference, the point of the log originally in contact with the ground comes to the top; then as the log rolls on another half-circumference, the point goes back to the bottom (its trajectory is a "cycloid"). It is as if this point traveled from a position below the object to a position on top of the object and back to below the object: So it appears to pass *over* the object (see Fig. 9.2). In other words, *over* applies to that point only, though we talk about motion of the whole log.

In the case of the earth turning around its axis, each point of the sphere, except the points on the axis, describes a circle around the axis. *Around* constrains the trajectory of every such point.

Consider also

The sea rushed onto the sand.

The water rose up to the roof.

In the first sentence, each wave (a part of a superficial layer of the sea) follows a path constrained by *onto the sand*; in the second, every part of water follows a path leading *up to the roof*. The prepositional phrases then describe the motions of each part (however defined) of the Figure object. Note that this *is* a schematization: The actual motions of the sea and of the rising water are far from so orderly. But it is not idealization of sea and water to a point.

In conclusion:

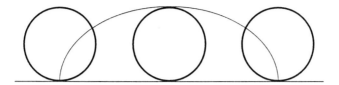

He rolled the log over

FIG. 9.2. Trajectory of a point on the circumference of a log in "He rolled the log over."

1. The primary meanings of the motion prepositions are all cast in terms of motion of a point.
2. Most often, such predicates are adapted to the motion of a three-dimensional object by reducing the motion to that of the object's centroid (abstracting the rotation components). This is effectively idealization to a point, and it follows in part from limitations in spatial cognition.
3. There are a few exceptions to this pattern: idiomatic ones ("roll over") and cases where the preposition applies to all or most points (or parts) of an extended Figure.

3. NAVIGATION AND COGNITIVE MAPS

So far, we have considered the use of motion prepositions from the point of view of perception, but prepositions are not uniquely, or even primarily for recording perceptual experience; importantly, they serve to describe action, specifically possible navigation paths. How to go from here to there is a central concern of human beings.

Navigation in large-scale spaces is guided by *cognitive maps* (Kuipers, 1983). These are unlike real paper maps; they are fragmented and distorted, but what is relevant here is that their major components are landmarks and routes (lines of travel between landmarks), which are represented respectively by points and lines. Moreover, a moving Figure in the context of a cognitive map is represented as a point and its trajectory as a line. The prepositions are frequently called upon to describe the location of punctual landmarks, linear routes, and paths of punctual moving objects in a cognitive map. So the sentences

Penny is at the market.

Penny walked to the market.

This street goes to the market.

refer to locations, trajectories, and pathways within a cognitive map. There is much linguistic evidence that one central sense of *at* is "coincidence of a movable point object with a point place in a cognitive map." For instance, *at* cannot be used to describe location in small-scale spaces:[4]

*The ashtray is at the bottle.
*The wastebasket is at the TV.

and

Jack is at the supermarket.

is typically infelicitous if the speaker herself is in the supermarket, because a space that surrounds you cannot be seen as a point; representing a fixed object as a point requires seeing it from a distance. In "I am at the supermarket." the speaker may be *in* the supermarket, but the sentence evokes a context where he is on the phone and taking the point of view of an auditor located at a distant place. Consider also

?*She is at the garden.
She is at the community garden.

The first is odd, as the vantage point evoked is from the house adjoining the garden; but the second is fine—the vantage point can be at a place distant from the community garden. The overall geometric context evoked is a cognitive map with landmarks and Figures represented as points.

So some idealizations are clearly grounded in spatial representations constructed for nonlinguistic purposes, namely in cognitive maps. The most basic sense of *at* is defined by reference to a cognitive map, and involves points as Figure and Ground.

4. IDEALIZATION TO A POINT AND DISTANCE BETWEEN FIGURE AND GROUND

When are idealizations of the Figure and/or Ground to a point appropriate for the other location prepositions? The answer depends on the sense of the preposition considered, but also sometimes on particular aspects of the referent situation, for example, the distance between Figure

---

[4]One sense of *at* entails, besides close proximity, a canonical interaction between a human Figure and the Ground: "Jane is at the desk/window." It can be extended to some inanimate Figures (e.g., "the chair at the desk"); but there is no general sense of *at* meaning "close proximity" (for details, see Herskovits, 1986).

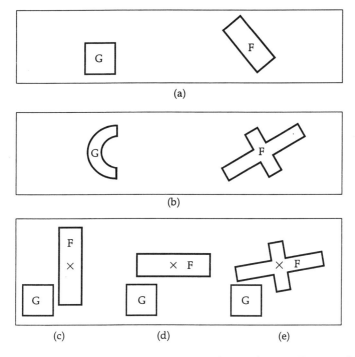

FIG. 9.3. Examples of the influence of the distance between Figure and Ground on the use of spatial prepositions.

and Ground. The "projective" prepositions (*to the right, above, behind*, etc.), for instance, exhibit a dichotomy: When Figure and Ground are far apart, they are viewed as points; when close, their shape and precise relative placement matter. This should come as no surprise. So in Figs. 9.3a and 9.3b we can decide that "F is to the right of G" by approximating F and G to points—their shape and orientation are irrelevant. But if Figure and Ground are close together (Figs. 9.3c through 9.3e), shape and exact relative placement matter: The rectangle in 9.3c is "to the right of G," but the rectangle in 9.3d and the cross in 9.3e are *not*, though the location of the centroid of F is the same in all three cases.

5. CONCLUSIONS

Schematization is a finely modulated process, dependent on many cognitive and linguistic variables. Table 9.2 contains a list of the elementary functions used in mapping real objects onto their schematic representation (for a precise description of these functions, see Herskovits, 1986, 1997).

TABLE 9.2
Schematization or Object Geometry Selection Functions

---

(1) idealizations to a:
- point
- line
- surface
- plane
- ribbon
(2) Gestalt processes:
- linear grouping (yields a two- or three-dimensional linear object)
- two- and three-dimensional grouping (yields an area or volume)
- completed enclosure
- normalized shape
(3) selections of axes and directions:
- model axis
- principal reference axis
- associated frame of reference
- direction of motion
- direction of texture
- direction of maximum slope of surface
(4) projections:
- projection on layout plane
- projection on plane of view
(5) part selections (triggered by high salience of the part):
- three-dimensional part
- oriented free top surface
- base

The applicable schematization can be the result of composing several elementary functions; so with "Mel walked across the streaming crowd." three such functions apply in succession: First, the many individual entities making up the crowd are turned into a volume bound by the "contour" of the crowd; second, this volume is projected on the layout plane to an area; third, a directionality—the common direction of motion of the crowd's members—is assigned to that area. Thus, the selection restrictions for one sense of *across* (an area with an intrinsic directionality) are satisfied, and Mel's path can be computed—it is in the ground plane, orthogonal to the area's directionality.

Sometimes, schematization reduces to the application of the trivial schematization function "identity." Then the full object shape is taken into account.

There are two important conclusions to draw from this look at schematization. The first concerns a hypothesis put forward by Landau and Jackendoff (1993), which states that the preposition and noun systems access distinct modules of the brain. There is evidence that object identification and object localization are performed by separate neural subsystems—the "what" and "where" systems (Ungerleider & Mishkin, 1982). Landau and Jackendoff proposed that the preposition system accesses only the encoding produced by the "where" system, and the noun system only the encoding produced by the "what" system. In support of this, they first claimed that we use detailed geometric properties of the objects when naming them (with nouns); but coarse representations—as points, lines (axial structure), and blobs—when locating them (with the help of a preposition). If "one kind of object description gets through the interface between spatial representations and language for naming (the 'whats'), and another kind of object descriptions does so for locating (the 'wheres')" (p. 257), then this difference between nouns and prepositions follows: The "what" system represents fine distinctions of shape; the "where" system represents objects only as place markers of roughly specified shapes.

But it is not the case that objects referred to in the context of a preposition are necessarily represented as points, lines, or blobs—one may need to know their shape and precise placement to decide upon the applicability of a preposition, even if the selection restrictions of the preposition do not actually specify the shapes of Figure and Ground. Landau and Jackendoff confused the selection restrictions (which may be in terms of points, lines, or blobs) with the knowledge of object shape needed for the preposition's appropriate use. Thus their hypothesis must be rejected.

The second important conclusion concerns artificial intelligence: Models that rely on some abstract representation of "world knowledge," ignoring schematization and its complex dependence on the perceptual, conceptual, and spatial cognitive systems, are bound to be limited. Research on spatial expressions must take the next step, toward the design of multimodal systems.

REFERENCES

Brugman, C. (1988). *The story of over: Polysemy, semantics, and the structure of the lexicon.* New York: Garland.

Herskovits, A. (1986). *Language and spatial cognition: An interdisciplinary study of the prepositions in English.* Cambridge, England: Cambridge University Press.

Herskovits, A. (1997). Language, spatial cognition, and vision. In O. Stock (Ed.), *Temporal and spatial reasoning.* Boston: Kluwer.

Kuipers, B. (1983). The cognitive map: Could it have been any other way? In H. Acredolo & L. Acredolo (Eds.), *Spatial orientation: Theory, research, and application* (pp. 345–359). New York: Plenum.

Landau, B., & Jackendoff, R. (1993). "What" and "where" in spatial language and spatial cognition. *Behavioral and Brain Sciences, 16,* 217–265.

Marr, D. (1979). Representing and computing visual information. In P. H. Winston & R. H. Brown (Eds.), *Artificial intelligence: An MIT perspective* (Vol. 2, pp. 17–82). Cambridge, MA: MIT Press.

Talmy, L. (1983). How language structures space. In H. Acredolo & L. Acredolo (Eds.), *Spatial orientation: Theory, research, and application* (pp. 225–282). New York: Plenum.
Ullman, S. (1985). Visual routines. In S. Pinker (Ed.), *Visual cognition* (pp. 96–159). Cambridge, MA: MIT Press.
Ungerleider, L. G., & Mishkin, M. (1982). Two cortical visual systems. In D. J. Ingle, M. A. Goodale, & R. J. W. Mansfield (Eds.), *Analysis of visual behavior* (pp. 549–579). Cambridge, MA: MIT Press.

10

Toward the Simulation
of Spatial Mental Images
Using the Voronoï Model

Geoffrey Edwards
Bernard Moulin
Université Laval, Quebec, Canada

We aim at developing software agents that are able to simulate the mechanisms involved when manipulating spatial mental images and reasoning about them. In order to simulate the way people usually reason about space, we need a model of space that allows qualitative reasoning. The Voronoï model of space is a good candidate for such a requirement. It provides an unambiguous way of defining neighbors based on relative proximity, and of constructing topological networks that encapsulate the spatial relationships between objects. In this chapter we show how various spatial prepositions can be mapped onto topological configurations in the Voronoï model. We show how we can develop a Voronoï agent that has the responsibility for maintaining a spatial database, for interpreting queries and commands expressed with prepositional descriptors, and for reasoning about the spatial relationships that result from these interpretations. In a multiagent system, the Voronoï agents are the only agents that handle spatial information. Other agents send queries to the Voronoï agents in order to get any topological or metric information characterizing the spatial relationships of the objects that are part of the space region covered by the Voronoï model. The Voronoï agents and their databases simulate a cognitive map whose information can be accessed by other agents. In addition, the information contained in these cognitive maps can be displayed to the user using familiar cartographic conventions.

1. INTRODUCTION

Understanding spatial knowledge is a complex cognitive activity involving a variety of inference tasks that require the same level of understanding as is assumed in functions supported by geographic information
systems (GIS) (Golledge, 1992). Although much of the research on cognitive mapping has shown that configurational-level understanding exists
in human beings, the nature of this type of knowledge is still poorly
understood. Several researchers have worked on *cognitive maps* as a basis
for representing configurational knowledge (Denis, 1989; Golledge & Zannaras, 1973; Gould & White, 1974). Researchers have tried to compare
cognitive and objective maps and to explain various aspects of configurational knowledge understanding as a function of neurobiological development (Golledge, 1992). The notions of mental model and mental
image have been also proposed as key representation schemes in cognitive
psychology (Johnson-Laird, 1983) and in linguistics (Langacker, 1991).
Hence, people appear to rely on mental imagery when they understand
a discourse as well as when they organize their ideas before generating
a discourse.

In this research we adopt Langacker's (1991) claim that "semantic
structures are characterized relative to *cognitive domains,* where a domain
can be any sort of conceptualization: a perceptual experience, a concept,
a conceptual system, an elaborate knowledge system, etc." (p. 3). When
a person reads a text or listens to a story, she imagines the corresponding
situations and projects on her "mental screen" the scenes in which characters interact. Thanks to the power of imagination, texts or discourses
become "animated products" evoking for the reader or listener real or
imaginary scenes that can be memorized, used to check the plausibility
of the narrator's claims or to reason about future outcomes. Our approach
has common grounds with Schirra's (1992) analysis of how verbal and
visual spaces are related in everyday life such as when radio reporters
describe sports scenes for the benefit of their auditors.

In the project VASSAR (Voronoï Agent for Spatial Simulation and
Reasoning) we aim at developing software agents that are able to simulate
the mechanisms involved when manipulating spatial mental images and
reasoning about them. These agents will be used for a variety of tasks,
including natural language interface, spatial analysis, reasoning, and way-
finding problems. In the coming years software agents (ACM, 1994) will
become more prevalent in everyday life. They will undertake routine
activities such as searching for information on networks, making reser-
vations, or ordering goods for their users. They will also need to provide
advice and information in a form that suits their users' cognitive abilities.
Understanding and explaining spatial knowledge will be a much needed

activity for different kinds of software agents. One of the principal questions of interest is whether it is possible to simulate such an abstract thing as a mental image, to reason with it and to display it to a user in the form of a concrete image.

2. THE VORONOÏ MODEL OF SPACE

First, let us recognize that representing and reasoning about spatial knowledge is a complex activity that applies, from a linguistic point of view, at three distinct levels: a *geometrical level* where topological, metric, and projective properties are dealt with; a *functional level* where we consider the specific properties of spatial entities deriving from their functions in space; and a *pragmatic level*, which gathers the underlying principles that people use in order to discard wrong relations or to deduce more information than can be actually found in a text (Aunargue, 1992; Aunargue & Vieu, 1994). In this research we assume that mental images are well suited to represent knowledge contained at the geometrical level, whereas the functional and pragmatic levels are best analyzed using symbolic knowledge adapted to reasoning activities. Because of the complexity of spatial knowledge representation and reasoning, we restrict our study to two-dimensional spatial information such as geographic data. We are particularly interested in how one may abstract information from this level in a form useful for the functional and pragmatic levels. We aim at developing mechanisms that will enable software agents to manipulate cognitive maps, in order to carry out these tasks.

In everyday life, people frequently use imprecise or fuzzy quantities when describing spatial information. Hence, in order to simulate the way people usually reason about space, we need a model of space that allows qualitative reasoning. This need becomes obvious when considering commonplace words and expressions such as "the next such and such . . . ," "the closest such and such . . . ," "between X and Y." Consider, for example, sentences like "this shop is next to the city hall," "that school is close to the river." We need a model that allows us to express qualitative values and relationships and to incorporate space in such an expressive structure. In addition, we need a model of space that can dynamically evolve when changes occur in the scene as observed by the software agent.

The Voronoï model of space (Okabe, Boots, & Sugihara, 1992) meets these requirements (Edwards, 1993; Gold, 1994). Each elementary map object (a point or a line segment, e.g.) is embedded in a *tile* (also called a *Voronoï region*), which is the region of space closest to the given object than any other object in the space. Voronoï regions can be determined for any arbitrary shape. They are commonly generated around points and line segments and recent work has extended them to curves and faces. The set of Voronoï regions for a set of objects in space is called the *Voronoï diagram* for the space and objects. It is also called a Dirichlet tesselation and the Voronoï regions are alternatively called Thiessen polygons. If the existence of a Voronoï boundary between two given objects can be established, then these objects are said to be neighbors, and we can say that they are adjacent. If the adjacency relations are represented by a line segment connecting the objects, then the set of all such line segments will form a Delaunay triangulation (for a set of points), called the dual of the Voronoï diagram. We call a *Voronoï model of space* the set of objects (half-lines and points) in two-dimensional space, their associated Voronoï regions, and the dual which contains the information on adjacencies. Objects are specified as collections of half-lines and points. We use the term *half-line* to denote a side of a line or, alternatively, a line and an orientation. All lines are composed of pairs of half-lines, although it is conceivable that a single half-line might sometimes be modeled.

The Voronoï model differs from the more commonly used raster and vector models[1] in many fundamental ways. Both raster and vector models are coordinate-based models or fully metric models of space, albeit with different metrics. The Voronoï model is not a fully metric model of space in that coordinates are not needed to determine adjacency relationships. The Voronoï model of space provides an unambiguous way of defining neighbors based on relative proximity, and of constructing topological networks that encapsulate the spatial relationships between objects. The Voronoï model of space is also fundamentally dynamic because no object may move without several other objects knowing it has done so. Object motion can be defined in terms of topological changes (changes of adjacency relations between objects) as well as relative changes in area. The Voronoï model is hierarchical in the sense that space can be nested into increasing levels of detail, each of which is embedded in the previous level. Objects and space are intimately connected. The nested and tiled nature of Voronoï space allows us to effectively remove any piece of a map and replace it with a simplified (i.e., generalized) version of the object at a lower level of detail. This is analogous to the nominalization process in natural language.

---

[1] In the *raster model* space is broken into regular discrete units called *cells*. Cells may be grouped into agglomerates called raster objects. Neighboring *raster objects* can be defined as those objects that share boundaries between outer cells. Topology is implicitly represented in this model via cell adjacencies, but no formal structure for expressing these relations is usually created in raster implementations. In the *vector model* space is defined by a Euclidian metric and an appropriate coordinate system. Objects are defined as collections of points, line segments, and polygons. Adjacencies can be defined only when two objects intersect. Neighborhood relations may be defined additionally via a metric relation.

Currently, our research partner has developed a library of functions that can be used to create Voronoï diagrams composed of points and line segments (Gold, 1994). This library[2] is currently applied to typical cartographic applications in GIS. The library may also be used to simulate many linguistic operators for handling spatial information.

3. A VORONOÏ PROCESSOR FOR SPATIAL EXPRESSIONS

In everyday life human beings usually rely on vague expressions to describe spatial situations or to formulate spatial queries. Most of the usual spatial expressions are based on some property of adjacency, neighborhood, closeness, or contact between objects. Such expressions are not easily transformed into queries to a database structured using a coordinate-based model or a fully metric model of space like the raster and vector models (Frank, 1992).

In contrast, the Voronoï library can be used directly to define map operators to implement a large number of linguistic concepts of space such as "near," "between," "among,"[3] and so on. For example, one of the query operators on the Voronoï data structure allows one to extract a quantity called StolenArea. This consists of the area of the Voronoï tile that would be gained by a given neighbor if the generating object were removed from the database. The quantity StolenArea was developed as a weight for interpolation problems (Gold, 1989), but can be used as a grading function for linguistic concepts. Hence, in Fig. 10.1, we see an example of a point query object (Q), which in (a) is treated as "far" from the pair of points labelled R1 and R2,[4] whereas in (b) it is considered to be "nearby." It is furthermore possible to compute a numerical value expressing this notion of proximity, which consists of the sum of the areas stolen by the query point from the tiles belonging to points R1 and R2,

---

[2]There are different types of functions. Some functions are used to create the Voronoï diagram (SetFrame, AddPoint, MovePoint, AddLine, JoinPoints, DeleteObject, etc.). Other functions are used to query the Voronoï structure, such as NearestObject, Neighbors, Trace, Clip, BufferZone, PolygonShade, StolenArea, and so forth. These functions are primitives that are used to manipulate the elements contained in the Voronoï data structure: points, line segments, and the triangles linking the objects.

[3]Let us note that to express most of these spatial expressions, we need to specify (at least implicitly) a query object that represents a specific location where the user (or an object pointed to by the user) is situated. For instance, in the sentence "Is there a cafeteria far from here?" *here* refers to the user's location.

[4]Our use of reference objects, a query point, and a test object is consistent with, if not identical to, existing approaches within the literature (e.g., Vandeloise, 1986).

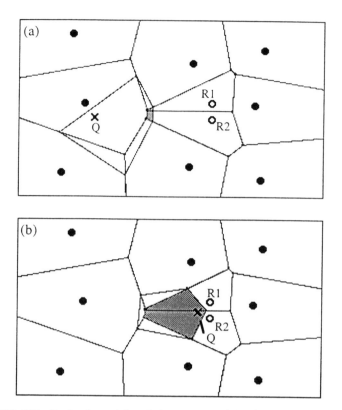

FIG. 10.1. Realizations of the relative concepts of (a) "far" and (b) "near" for a query point Q with respect to the pair of reference points R1 and R2, using the Voronoï StolenArea operation.

divided by the total area occupied by the query point. The StolenArea is a standard query supported by the Voronoï library.

This measure of proximity is computed relative to the presence of other points around the object, and hence interprets "near" to mean "nearer than the surrounding points." A different measure of proximity was discussed in Edwards (1993), permitting an interpretation of near and far in terms of the size of the reference object. Both methods are of interest.

In Fig. 10.1b, the query point could be considered to be "between" the reference points R1 and R2, because the area stolen from point R1 by the query point is similar to the area stolen from point R2. In order to be "directly between" points R1 and R2, however, the query point would have to steal area from (or, equivalently, include in its set of neighbors) the points on opposite sides of the reference pair (Fig. 10.2a). For pairs of line segments, the equivalent condition is that the query point be cut off from the neighbors on each end of the line segment pair (Fig. 10.2b).

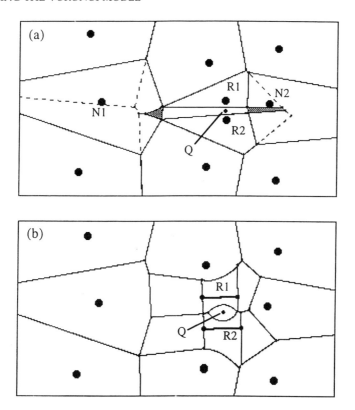

FIG. 10.2. Examples of the Voronoï configuration for the preposition *between*, (a) for point reference objects, and (b) for line reference objects. The dashed lines indicate the Voronoï boundaries if points N1 and N2 are removed.

The concept of "behind" versus "in front of" can be handled in a similar manner. In Fig. 10.3, we use the same frame of points as in Fig. 10.1. We define the reference object R to be fixed in space, the Query point Q to be displacable, and the Test object T also to be displacable. In Fig. 10.3, the test point T is located "on the other side" of point R from the query point Q. The logic of this situation is that both the query point and the test point steal area from the reference point, but they do not steal area from (and are not neighbors of) each other. This can be expressed in formal terms:

$$\text{"behind"} = ((Q,R),(R,T),\neg(Q,T))$$

where (x,y) indicates that the object pair in the parentheses are neighbors of each other and $\neg$ indicates negation.

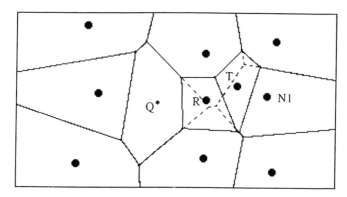

FIG. 10.3. Example of the Voronoï configuration where the test point T is considered to be "behind" the reference point R, as seen by the query point Q. The dashed lines indicate the Voronoï boundaries if points R and T are removed.

The concept "in front of" is represented when the query point and the target point become neighbors, whereas the target point keeps the reference point as a neighbor. Hence the relationship can be expressed as:

$$\text{"in front of"} = ((Q,T),(T,R),\neg(Q,R)).$$

In a similar way, it is possible the determine a configuration for each spatially referenced preposition. Indeed, and this is the reason this approach is so powerful, *each preposition is mapped to a different topological configuration*, where topology is used in the Voronoï sense of indicating the existence of an adjacency relation between Voronoï regions. In addition, for each topological configuration (and its corresponding preposition), one or more grading functions may be defined. Table 10.1 shows a list of topological configurations for the following static spatial prepositions: *far from, near to, behind, in front of, to the side of, to the right side of, to the left side of, between, among, amid, inside,* and *outside.* For each preposition, the situation variables are defined, the topology is given in the formal sense as described earlier, and the same topological information is given in the form of a visual graph. Furthermore, a grading function is proposed for each preposition. The grading functions, defined in terms of ratios of Voronoï areas, vary between extreme values of 0 and 1. These grading functions are related to modifier terms applied to the prepositions. Characteristic values of these modified prepositions are indicated for each end of the graded range.

Furthermore, Table 10.1 contains an additional subclass of prepositions. Indeed, prepositions such as *near* or *inside* do not require a privileged point of view, whereas prepositions such as *behind* and *to the side of* implicitly require an orientation. In the Voronoï model, this orientation is provided

TABLE 10.1
Static Spatial Prepositions

Preposition	Situation[a]	Topology[b]	Topological Configuration	Grading Function[c]	Graded Preposition
proximity	$(Q{\rightarrow}R)$	(Q,R)	Q—R	$\dfrac{SA(R{\rightarrow}Q)}{A(Q)}$	Near - Far
between	$(Q{\rightarrow}(R1,R2))$	$((Q,R1),(Q,R2))$		$\dfrac{SA(R1{\rightarrow}Q)-SA(R2{\rightarrow}Q)}{SA(R1{\rightarrow}Q)+SA(R2{\rightarrow}Q)}$	perhaps between - directly between
behind	$(Q{:}T{\rightarrow}R)$	$((Q,R),(R,T),\neg(Q,T))$		$\dfrac{SA(T{\rightarrow}Q,\neg R)}{A(Q)}$	perhaps behind - directly behind
in front of	$(Q{:}T{\rightarrow}R)$	$((Q,T),(T,R),\neg(Q,R))$		$\dfrac{SA(R{\rightarrow}Q,\neg T)}{A(Q)}$	perhaps in front of - directly in front of
to the side of	$(Q{:}T{\rightarrow}R)$	$((Q,R),(T,R),(T,Q))$		$\dfrac{A(Q)-SA(T{\rightarrow}Q)}{A(Q)}$	perhaps to the side - directly to the side
to the right side of	$(Qr{:}T{\rightarrow}R)$	$((Qr,R),(T,R),(T,Qr))$		$\dfrac{A(Qr)-SA(T{\rightarrow}Qr)}{A(Qr)}$	perhaps right - directly right
to the left side of	$(Ql{:}T{\rightarrow}R)$	$((Ql,R),(T,R),(T,Ql))$		$\dfrac{A(Ql)-SA(T{\rightarrow}Ql)}{A(Ql)}$	perhaps left - directly left
inside of	$(Q{\rightarrow}R)$	$((Q,Ri(j))j=1,\ldots,n))$		$\dfrac{Sum(SA(Ri{\rightarrow}Q))}{A(Q)}$	partially inside - fully inside
outside of	$(Q{\rightarrow}R)$	$((Q,Ro(j))j=1,\ldots,m))$		$\dfrac{A(Q)-Sum(SA(Ri{\rightarrow}Q))}{A(Q)}$	partially outside - fully outside
among	$(Q{\rightarrow}(R1,R2,\ldots,Rm)$	$((Q,R(j))j=1,\ldots,m))$		$\dfrac{A(Q)-Sum(SA(R(j){\rightarrow}Q))}{A(Q)}$	partially among - fully among
amid	$(Q{\rightarrow}(R1,R2,\ldots,Rm)$	$((Q,R(j))j=1,\ldots,m))$		$A(Q)/m-Ave(SA(R(j){\rightarrow}Q))$	perhaps amid - directly amid

[a]The symbol $\rightarrow$ means "with respect to"; "Q:T$\rightarrow$R" means that the axis Q-R defines the orientation, with Q the viewpoint; Q = Query object; R = Fixed Reference object; T = Movable test object.

[b]Ri = Reference objects labeled as Interiors; Ro = Reference objects labeled as exteriors; N = A neighboring object; (p,q) means that object p is adjacent to object q; ¬(p,q) means that object p is not adjacent to object q. Qr (respectively Ql) means the part of the Voronoï zone of Q situated on the right-hand side (resp. left-hand side) of object Q.

[c]A (p) means the Voronoï area of object p; SA(p$\rightarrow$q) means the Voronoï area of object p stolen by q.

171

by the relative position of the query point with respect to the reference object. Table 10.2 shows the topological configurations for dynamic spatial prepositions, such as *beyond, toward, through,* and *around.* These are dynamic in the sense that they imply the movement of the query point, and hence must be represented as a sequence of configurations. The dynamic properties of the Voronoï model of space allow us to realize dynamic scenarios based on these sequences. Hence, for example, the preposition *around* is represented as the process by which the query point acquires as neighbor to each of the reference point's neighbors in turn. This topological definition is equivalent to modeling the movement of the query point around the reference point in the Voronoï data structure.

In Table 10.2, we indicate neighboring objects or points that are not part of the situation definition with the symbol N. The prepositions *inside* and *outside* are determined as a function of the query point being surrounded by points or objects that are labeled *inside* and *outside*. This is not as arbitrary as it sounds. In fact, because lines have sidedness, the space neighboring a set of lines that form a closed box (or not quite closed—closure is not required) will be divided in a natural fashion into "interior" and "exterior" regions (Fig. 10.4). This situation can also be simulated with a double series of points (Fig. 10.5), with the inside set labeled *interior* and the outside set labeled *exterior* points.

Left and right are also inherently defined when the query object is a line segment (as are front and back), but can be simulated with point objects by dividing the Voronoï zone of the query object into two pieces, using the reference object to define the orientation (Fig. 10.6). It should also be noted that not all these topology configurations are appropriate in the presence of line segments. Those shown are derived for proximity to points, and may need to be modified to handle line segments appropriately. This is seen more clearly in the example shown in Fig. 10.2 for the case of *between.*

Throughout the discussion we have referred to points and more complex objects as if they were interchangeable. This is because the Voronoï model of space allows us to generalize complex objects as if they were points (Edwards, 1993). The use of StolenArea-based grading functions is also based on this situation. Distance-based grading functions could be used, in principle, but they are not easy to define in the presence of complex objects. Other prepositional locutions are possible by combining those shown in Tables 10.1 and 10.2. Hence, for example, "to go into a space" may consist of going "between" or "through" some reference objects and arriving in a space labeled *inside.*

An additional notion that we are required to assert in order to be able to correctly characterize linguistically described situations is the idea of an abstraction operator. Indeed, because the Voronoï operates with respect

TABLE 10.2
Dynamic Spatial Prepositions

Preposition	Situation	Number of Configurations	Topology	Topological Configuration
toward	$(Q \to R)$	3	$\neg(Q,R)$ (Q,R) (Q,R) + decreasing proximity	
through	$(Q \to (R1,R2))$	3	$(Q,N1),(Q,R1),(Q,R2)$ $(Q,N1),(Q,R1),(Q,R2),(Q,N2)$ $(Q,R1),(Q,R2),(Q,N2)$	
beyond	$(Q{:}T \to R)$	2	$(Q,R), (R,T),\neg(Q,T)$ $(Q,R),\neg(R,T),\neg(Q,T)$	
around	$(Q \to R)$	n = number of neighbours of R	$(R,Q),\neg(R,N1), (R,N2),\ldots, (R,Nn)$ $(Q,N2),(N2,\ldots),\ldots,(\ldots,Nn),(Nn,Q))$ $(R,Q), (R,N1),\neg(R,N2),\ldots, (R,Nn)$ $(Nn,N1),(N1,Q)\ldots, (\ldots,Nn)$ .... $(R,Q), (R,N1), (R,N2),\ldots,\neg(R,Nn)$ $(N1,N2), (N2,\ldots),\ldots, (Q,N1)$	

Note. For a description of the symbols, see Table 10.1.

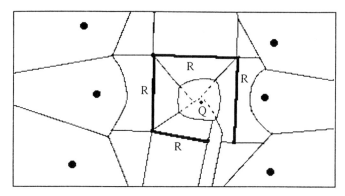

FIG. 10.4. Example of the Voronoï configuration for the preposition *inside*, where the query point Q is considered to be inside the box R. The dashed lines indicate the Voronoï boundaries if the point Q is removed. Notice how the region interior to the line segments is well defined, even though there is an opening in the box. The entry way contributes very little area to the query point, although it is a neighbor.

to the objects present in its "space," it is necessary to be able to remove objects that are not considered part of any situation, or to handle only the few objects directly affected by the description. Hence, we must be able to abstract subsets of objects to a buffer space in which the Voronoï model allows us to map topological configurations into propositional descriptions. This constitutes both an advantage and a disadvantage. The advantage is that, whereas the Voronoï operators ostensibly work only on objects that are Voronoï neighbors, the abstraction mechanisms allows us to deal with a wide variety of circumstances under which objects *are*

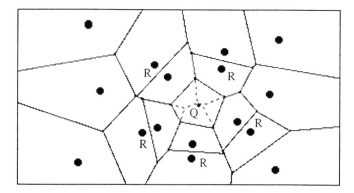

FIG. 10.5. Example of the Voronoï configuration for the preposition *inside*, where the query point Q is inside a double ring of points R. The dashed lines indicate the Voronoï boundaries if the point Q is removed. Note that the query point Q steals area from the interior ring of points only.

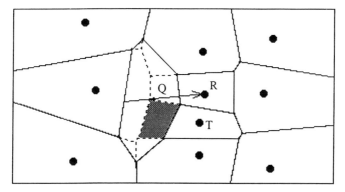

FIG. 10.6. Example of the Voronoï configuration for the preposition *to the right side of*. The dashed lines indicate the Voronoï boundaries if the point Q is removed. T steals a sizable piece of the right side of Q, and hence one can deduce that T is almost directly to the right of R as seen by Q.

not immediate neighbors. This is actually similar to the way linguistic/ perceptual operators work. We may say (Fig. 10.3) that object N1 is behind object R from the point of view of Q, just as we may say object T is behind object R from the same point of view. When we do this, however, we have visualized the situation as containing only the points R and N1 (and the other neighbors) in the first case, and only the points R and T in the second case. Hence, we have selected different object subsets when characterizing these situations. In other words, we have changed the context. The disadvantage of this, of course, is that the way the abstraction operator works must be carefully defined, in order to ensure that the "correct" object subsets are chosen. The abstraction operator will largely operate symbolically or thematically, although it may sometimes by evoked by the use of demonstratives ("this one over here," "that one over there"), by the geometric properties of objects ("the circular building"), by spatial prepostions operating on preexisting object subsets ("the building between Building A and Building B"), or by any combination of these ("that little building just behind Building A, over here").

Given these elements, spatial information in the form of prepositions can be converted into topological configurations. From this, it should be possible to be able to reason about different spatial relationships and, indeed, deduce new relationships from those that are stated explicitly. For example, if we state that Q is near R, T is near R, and Q is near T, then we may conclude from our previous definitions that T is on one side of R with respect to Q. The one-to-one correspondance between different prepositions and different topological configurations ensures that such a spatial reasoning system would have great expressive power. The process of spatial reasoning is rendered more complex, however, by the abstraction operator described in the previous paragraph. Thus, it will be necessary not only to compare and combine spatial relations, but also to reason about the context of the relations—that is, the inclusions and exclusions of objects within the subsets used to process the spatial relations. This is actually a powerful feature of the model proposed, because it defines context in a very specific way, and one that is compatible with its use in natural language.

The approach to natural language understanding just described deals only with spatial prepositions, so-called closed-class lexical elements, and not with spatially referenced adjectives or nouns, or open-class lexical elements. Hence, the methodology outlined is not complete. Furthermore, we have excluded from the discussion spatial prepositions that refer to the vertical dimension. Some of these were discussed in Edwards (1993). Furthermore, an extensive study of near-field effects in the presence of line segments may yield additional linguistic operators. For example, the interpretations of *at* and *next to* appear to correspond to additional topological configurations.

In Edwards (1993), it was shown that the spatial relations expressed in the Voronoï model of space have the same four neutrality properties as those found in natural language—magnitude neutrality, shape neutrality, closure neutrality, and continuity neutrality (Talmy, 1983). This means that under changes in any of these properties, the applicability of the prepositions is unchanged. We saw an example of the Voronoï's insensitivity to closure in Fig. 10.4, whereas its insensitivity to continuity is shown in Fig. 10.5. Furthermore, Gold (1992) emphasized the important links that exist between human perception and the Voronoï model. Hence the "fit" between the Voronoï model and natural language appears to be remarkably good.

In summary, the topological manner in which spatial relations are expressed via the Voronoï diagram and its dual, the Delaunay triangulation, provides a natural framework for developing a spatial logic that can be used to reason about objects in space. The method does not require any recourse to metrical methods of classifying proximity relations. Most prepositions are mapped onto a distinct topological configuration. Only prepositions such as *near* and *far* are obtained by grading a common topological configuration.

4. PROCESSING SIMULATED COGNITIVE MAPS

The Voronoï library can be embedded into one or more specialized Voronoï agents, which have the responsibility for maintaining a spatial database, for interpreting queries and commands expressed with prepositional descriptors, and for reasoning about the spatial relationships that result from these interpretations. The Voronoï agents play the role of specialized servers that can be queried by other agents when they need to access to the spatial information corresponding to a specific region of space, or to information about the spatial relations between objects in the database.

From a knowledge representation perspective, the Voronoï agents can be viewed as systems-manipulating cognitive maps for the benefit of other agents. Indeed, the Voronoï agents act as autonomous systems that provide a database representing a given space region and dynamically maintain its topological and metric properties. Whenever a change (such as an object's movement) occurs, a perception agent notifies the appropriate Voronoï agent, which dynamically updates its database and its topological proper-ties. Furthermore, a second Voronoï agent could maintain a data store of maps of possible spatial relations as they relate to prepositional structures, such as those shown in Tables 10.1 and 10.2, and could use these schemas to process requests from other agents regarding such relations. As an example let us consider a multiagent system responsible for orienting the persons who are unfamiliar with the campus configuration of Laval Uni-versity (Fig. 10.7). This application is built as a multiagent system (Moulin & Cloutier, 1994, 1995) composed of six specialized agents: a coordinating agent called Laval-Guide, an InteractionAgent, a Voronoï agent called Laval-Vagent, a NavigatorAgent, a DisplayAgent, and a second Voronoï agent called Vmap Construction Agent.

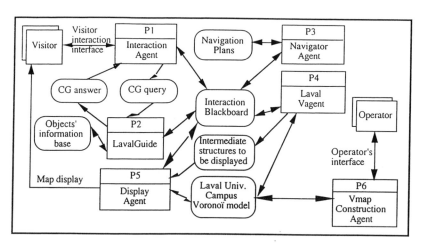

FIG. 10.7. An overview of the orientation multiagent system. Rectangles represent agents, round-cornered rectangles represent data or knowledge repositories, and arrows represent data or knowledge exchanges between agents and repositories. The double squares represent the human users of the system.

The coordinating agent Laval-Guide maintains a symbolic representation of all the relevant objects located on the campus: buildings and their doorways, streets and paths crossing the campus, different points of interest, and so on. The symbolic representation of an object uses a knowledge representation formalism such as that provided by conceptual graphs (Sowa, 1984) to record the object identifier, the object description, and information to be given to the visitor. In addition, Laval-Guide is responsible for coordinating the activities of the other agents composing the multiagent system. All of the spatial information characterizing the objects located on the Laval University campus is maintained by the Voronoï agent Laval-Vagent, which uses the same object identifiers as Laval-Guide. Laval-Vagent also maintains the cognitive schema, which allows it to translate linguistic structures such as prepositions into topological information expressing the spatial relationships between objects.

A visitor can express requests on a screen by instantiating query frames that are selected through a user-friendly interface (using menus, icons, dialog boxes, etc.). The visitor might also have access to pseudonatural language interfaces that allow the expression of spatial relationships, such as through the use of a visual language. The InteractionAgent processes these requests and transforms them into a semantic structure in the form of conceptual graphs that will be processed by LavalGuide. There are different categories of requests that a visitor can make: (a) information requests (i.e., What departments are located in building XX? What is the list of libraries and their areas of specialization? In which building is displayed exhibit EE?, etc.), (b) localization requests (i.e., Where is building XX? Is the entrance of building YY far from here? What is the nearest cafeteria? Is there a cafeteria next to building YY?, etc.), and (c) path-finding requests (i.e., How can I go to building XX? What is the best way to go from place PP to building YY?, etc.).

Depending on the category of a query, LavalGuide may have different options. For an information request, it may select the answer in its database and send the information to the InteractionAgent, which transforms it into a sentence form and displays it to the user. For a localization request (or any request involving spatial referencing), it may send the spatial part of the request to the Voronoï agent Laval-Vagent, which interprets the request. Interpreting consists of classifying the type of spatial information requested—direct coordinate information, information about spatial relationships that may be extracted directly from the database, or information that requires spatial inference or spatial analysis. The VAgent may also need to determine whether the request contains enough information to permit an unambiguous answer. After ensuring that the request is complete, Laval-Vagent will determine the response and express it either within intermediate graphic structures to be displayed on the visitor's screen, or as conceptual graphs that will be used by the InteractionAgent to formulate an answer in natural language. Control is passed back to LavalGuide, which asks the InteractionAgent to generate the answer in natural language and tells the DisplayAgent to illuminate the specific elements of the map that correspond to the visitor's request. LavalGuide would then pass control to the InteractionAgent, which would wait for the next visitor's request. For a path-finding request, LavalGuide would send a query to a NavigatorAgent, which possesses procedures to transform path-finding requests into queries sent to the Voronoï agent Laval-Vagent. When the solution is found, thanks to the cooperation of the NavigatorAgent and Laval-Vagent, it can again be displayed on a map by the DisplayAgent, or transformed into natural language by the InteractionAgent.

Another agent, called the VmapConstructor, may be used by the person in charge of the system in order to create and update the Voronoï database. This agent would have functionalities similar to those found in the cartographic system based on the Voronoï model presented in Gold (1994). New information about the localization of objects (exhibits, buildings, repairs, suggested paths, etc.) on the campus would be directly recorded into the Voronoï database using a direct manipulation interface that allows the operator to update the object network.

In this system the Voronoï agents are the only agents that handle spatial information. Other agents send queries to the Voronoï agents in order to get any topological or metric information characterizing the spatial relationships of the objects that are part of the space region covered by the Voronoï model. In addition, only the VmapConstructor Voronoï agent is required to update the baseline database, which it does using conventional GIS operations. Hence, there is no need to map linguistic expressions "downward" into an explicit spatial representation. The Voronoï agents and their databases simulate a cognitive map whose information can be accessed by other agents. In addition, the information contained in these cognitive maps can be displayed to the user using familiar cartographic conventions. A blackboard data structure (Nii, 1986) may be used to coordinate the interactions taking place between the various agents of the system. This blackboard (round-cornered rectangle in the center of Fig. 10.7) would be controlled by LavalGuide. Agents would post requests to other agents on the blackboard as well as partial solutions contributing to the current problem resolution.

On the map displayed on the screen, a user can point to objects to using a pointing device such as a mouse. The Voronoï data structure facilitates such queries, because pointing at the object itself is not necessary—the user need only point to the Voronoï region of the desired object. Such an operation corresponds to linguistic expressions using demonstratives like *this* or *that*, and may be processed by the Voronoï agent as such. When answering a user's query, the DisplayAgent could illuminate the relevant objects on the displayed map, while the InteractionAgent in cooperation with the Voronoï agent LavalVagent would formulate the answer using natural language expressions.

5. TRANSLATING CONCEPTUAL GRAPHS INTO VORONOÏ OPERATIONS

We indicated that the InteractionAgent translates a visitor's request into a conceptual graph (CG) form that is processed by LavalGuide who transforms it into queries to the Voronoï agent Laval-Vagent. In this section we show how those CG requests are transformed into operations on the Voronoï database. Here, we only analyze some localization requests. We examine how these requests are specified in a CG form, and how the CG requests are mapped to a query using the Voronoï function library. We express these CG requests using the linear form of conceptual graphs.[5] We use a set of spatial relations suggested by Willems (1993) to extend the expressive power of conceptual graphs on the basis of spatial

---

[5]Here, we briefly introduce the linear notation used in CG theory. For more details see Sowa (1984). There is also a graphical representation for CGs. Concepts are the representations of objects of the application domain. A concept is characterized by two elements: a type, which represents the set of all the occurrences of a given class (i.e., human, animal, etc.); and a referent, which represents a given occurrence of the class that is associated with the concept (i.e., John, Mary, etc.). A concept is specified between square brackets: [TYPE-NAME: referent]. An elementary link between two concepts is called a conceptual relation. Using linear notation, the conceptual relation is represented between brackets and is associated with concepts by means of arrows. Conceptual relations are usually binary. They often correspond to "semantic cases" (Fillmore, 1968) such as "agent" (AGNT), "object" (OBJ), "instrument" (INST), "patient" (PTNT), "theme" (THM), "characteristic" (CHRC), "point in time" (PTIM), "beneficiary" (BENF), "duration" (DUR), and so on. Conceptual graphs are finite, connected, and bipartite graphs, whose nodes are either concepts or conceptual relations. The CG theory enables us to specify embedded CG by the use of a concept of type PROPOSITION. For instance "Mary thinks John eats an apple" is expressed by:

> [PERSON: Mary] ← (PAT) ← [THINK] → (THM) →
> [PROPOSITION: [PERSON: John] ← (AGNT) ← [EAT] → (OBJ) → [APPLE]].

Sowa (1988) defined a "context" as a set of propositions, and a situation as "a state of affairs that occurs at a single place and time" (p. 8). For instance we can indicate the situation: "John eats an apple" by:

> [SITUATION: [PERSON: John] ← (AGNT) ← [EAT] → (OBJ) → [APPLE]].

primitives proposed by Jackendoff (1990). Here are some examples of such requests.

The request "Where is building X?" is translated into the conceptual graph (CG):[6]

$$[\text{BUILDING: X}] \rightarrow (\text{LOC}) \rightarrow [\text{PLACE: ?}]. \qquad \text{(CG1)}$$

where the question mark identifies the element that is the object of the request. This is a simple request that is directly mapped to a query to the Voronoï engine of the form of

$$\text{Display (Set-of-Points, Ident (X))}$$

which means "Display the set of points corresponding to the complex object with the identifier Ident (X)."

The request "Is the entrance of building X far from here?" is translated into a CG:

$$\begin{aligned} &[\text{ENTRANCE: \# }]\text{-} \qquad\qquad\qquad\qquad\qquad\qquad\quad \text{(CG2)}\\ &(\text{PART-OF}) \rightarrow [\text{BUILDING: X}]\\ &(\text{LOC}) \rightarrow [\text{PLACE:*}] \rightarrow (\text{FAR:?}) \rightarrow [\text{POSITION: MyPosition}]. \end{aligned}$$

In this request we can distinguish a purely semantic part introduced by the PART-OF relation, and a geometric part introduced by the LOC relation. This geometric part will be translated into a query to the Voronoï agent. In the concept POSITION, the referent MyPosition refers to the user's current position. This will be mapped to the Query point Q in the Voronoï database. The relation FAR is an abstract conceptual relation (similar to those proposed by Willems, 1993) that catches the meaning of the natural language expression. If we were to adopt a purely semantic approach, we would have to translate that relation into a set of primitive ones, which is not a simple endeavor. Instead, the relation will be translated into queries that take advantage of the topological operations on the Voronoï database. First, the identifier Ident (ent) of the entrance of building X should be retrieved by LavalGuide in its database. Then, if

---

[6]In the referent field of a concept, we can specify special characters. The symbols # and * correspond to the definite article *the* and to the indefinite article *a*, respectively. If there is no element in the referent field of a concept, this is equivalent to the indefinite article. Note that we extend Sowa's (1984) notation by allowing referents in relations, in order to specify a question mark indicating that the question focuses on the relation, as in the example sentence "Is the house or room the school?":

$$[\text{HOUSE: \#}] \rightarrow (\text{FAR: ?}) \rightarrow [\text{SCHOOL: \#}].$$

we call Re the object of the Voronoï database with the same identifier Ident (ent), the query to the Voronoï engine bears on the grading function SA (Re → Q) (see Table 10.1). After obtaining the value of the grading function from the Voronoï agent, LavalGuide is able to reason about it in order to determine whether the answer should be "Yes, it is far from here?" or "it is within a walking distance from here," or any other appropriate answer from the visitor's point of view.

The request "What is the nearest cafeteria?" cannot be directly translated into a CG because the answer requires first looking for all the cafeterias Ci such that

$$[\text{CAFETERIA: Ci}] \rightarrow (\text{LOC}) \rightarrow [\text{PLACE:}] \rightarrow [\text{POSITION: MyPosition}]$$

and passing their idents to the Voronoï Laval-Vagent. The answer to the request will then be furnished by the latter. Due to space limitations, we cannot present the full variety of requests that can be processed by the Voronoï agent. We hope that this presentation gives the reader a flavor of the possibilities of our approach.

6. CONCLUSIONS

We have outlined how the Voronoï model of space, especially as it has been implemented within a working Voronoï software library through the work of Gold (1994), could be used to improve the functionality of a multiagent system designed to provide spatial information about a given context. The Voronoï agents in the system handle all spatial operations, including building and modifying the spatial databases and processing spatial queries. The multiagent system uses topological primitives to interpret spatial prepositional structures and convert them into spatial configurations. These topological primitives form a kind of dictionary of spatial relationships, and should permit the multiagent system to reason "intelligently" about the spatial relations present in the database and to communicate the results of the reasoning process. The fact that the multiagent system uses such a dictionary allows it to determine whether or not an incoming query contains sufficient information to provide an unambiguous answer.

Although the Voronoï model of space only encodes explicit information about immediate neighbors, the use of an abstraction procedure allows it to handle a wide variety of circumstances where objects are not necessarily immediate neighbors. This is done by first using an abstraction operator to select object sets of interest that are moved into a temporary

Voronoï space for spatial processing. In this way it is possible to "filter" out intervening objects.

The Voronoï agent manipulates a cognitive map coded in a topological network, for which translation mechanisms exist for converting the map information "upward" into conceptual information in the form of map schema that are matched to spatial prepositions in natural language. In principle, the Voronoï agent could also be used to convert cognitive map schema derived from natural language "downward" into geometric information. For the application described, this is not presently necessary, and indeed much more work would need to be carried out to define how this should be done. Thus, the Voronoï agent serves as a bridge between geometry and logic. The schemas maintained by the Voronoï agent are clearly compatible with the schemas humans maintain in order to handle linguistic structures, although we cannot assert that they are the same. Circumstantial evidence suggests that the map is pretty good, but for machine processing of spatial information, we do not need a cognitive map identical to the one humans use, only one that is efficient at carrying out the tasks we wish to perform.

ACKNOWLEDGMENTS

This research was supported by the Natural Sciences and Engineering Research Council of Canada and the Association des industries forestières du Québec via the foundation of an Industrial Research Chair in Geomatics.

REFERENCES

ACM. (1994). Intelligent agents [Special issue]. *Communications of the ACM, 37*(17).

Aurnargue, M. (1992). A unified processing of orientation for internal and external localization. In M. Aunargue, A. Borillo, M. Borillo, & M. Bras (Eds.), *Semantics of time, space, movement and spatio-temporal reasoning* (pp. 39–52). Toulouse, France: IRIT.

Aurnargue, M., & Vieu, L. (1994). A three level approach to the semantics of space. In C. Zelinsky-Wibbelt (Ed.), *The semantics of prepositions: From mental processing to natural language processing.* Berlin: Mouton de Gruyter.

Denis, M. (1989). *Image and cognition.* Paris: Presses Universitaires de France.

Edwards, G. (1993). The Voronoï model and cultural space: Applications to the social sciences and humanities. In A. U. Frank & I. Campari (Eds.), *Spatial information theory—A theoretical basis for GIS. Lecture notes in computing sciences, 716* (pp. 202–214). Berlin: Springer-Verlag.

Fillmore, C. H. (1968). The case for case. In Bach & Harms (Eds.), *Universals in linguistic theory.* New York: Holt, Rinehart & Winston.

Frank, A. U. (1992). Qualitative spatial reasoning about distances and directions in geographic space. *Journal of Visual Languages and Computing, 3*, 343–371.

Gold, C. M. (1989). Surface interpolation, spatial adjacency and GIS. In J. Raper (Ed.), *Three dimensional applications in geographic information systems* (pp. 21–35). London: Taylor & Francis.

Gold, C. M. (1992). The meaning of "neighbour." In A. U. Frank, I. Campari, & U. Formentini (Eds.), *Theories and methods of spatio-temporal reasoning in geographic space. Lecture notes in computing science, 639* (pp. 220–235). Berlin: Springer-Verlag.

Gold, C. M. (1994). The interactive map. *Journal of the Geodetic Commission of the Netherlands, 40,* 121–128.

Golledge, R. G. (1992). Do people understand spatial concepts: The case of first-order primitives. In A. U. Frank, I. Campari, & U. Formentini (Eds.), *Theories and methods of spatio-temporal reasoning in geographic space. Lecture notes in computing science, 639* (pp. 1–21). Berlin: Springer-Verlag.

Golledge, R. G., & Zannaras, G. (1973). Cognitive approaches to the analysis of human spatial behaviour. In W. Ittelton (Ed.), *Environmental cognition* (pp. 59–94). New York: Seminar Press.

Gould, P., & White, R. (1974). *Mental maps.* Harmondsworth, England: Penguin.

Jackendoff, R. (1990). *Semantic structures.* Cambridge, MA: MIT Press.

Johnson-Laird, P. N. (1983). *Mental models.* Cambridge, England: Cambridge University Press.

Langacker, R. W. (1991). *Concept, image and symbol, the cognitive basis of grammar.* Berlin: Mouton de Gruyter.

Moulin, B., & Chaib-draa, B. (in press). Distributed artificial intelligence: An overview. In N. Jennings & G. O'Hare (Eds.), *Foundations of distributed artificial intelligence.* New York: Wiley.

Moulin, B., & Cloutier, L. (1994). Collaborative work based on multiagent architectures: A methodological perspective. In F. Aminzadeh & M. Jamshidi (Eds.), *Soft computing: Fuzzy logic, neural networks and distributed artificial intelligence* (pp. 261–296). Englewood Cliffs, NJ: Prentice-Hall.

Nii, H. P. (1986). Blackboard systems: The blackboard model of problem solving and the evolution of blackboard architectures (part I). *AI Magazine, 7*(2), 38–53.

Okabe, A., Boots, B., & Sugihara, K. (1992). *Spatial tessellations—Concepts and applications of Voronoï diagrams.* Chichester, England: Wiley.

Schirra, J. R. J. (1992). Connecting visual and verbal space. In M. Aunargue, A. Borillo, M. Borillo, & M. Bras (Eds.), *Semantics of time, space, movement and spatio-temporal reasoning* (pp. 105–121). Toulouse, France: IRIT.

Sowa, J. F. (1984). *Conceptual structures.* Reading, MA: Addison-Wesley.

Sowa, J. F. (1988). Conceptual graph notation. In J. W. Esch (Ed.), *Proceedings of the 3rd Annual Workshop on Conceptual Graphs, St. Paul, Minnesota.*

Talmy, L. (1983). How language structures space. In H. Pick & L. Acredolo (Eds.), *Spatial orientation: Theory, research and application* (pp. 225–282). New York: Plenum.

Vandeloise, C. (1986). *L'espace en Français* [Space as expressed in French]. Paris: Presses Universitaires de France.

Willems, M. (1993). A conceptual semantics ontology for conceptual graphs. In G. W. Mineau, B. Moulin, & J. F. Sowa (Eds.), *Conceptual graphs for knowledge representation. Lecture notes on artificial intelligence, 699* (pp. 312–327). Berlin: Springer-Verlag.

11

Generating "Mental Maps" From Route Descriptions

Lidia Fraczak
Université de Paris-Sud

Despite the fact that spatial expressions are an important topic in cognitive science, route descriptions, produced in everyday communication situations, have not been studied much so far. Yet they are a real challenge from a linguistic as well as a cognitive science point of view, as they involve specific forms of representation. This chapter deals with the generation of "mental maps" from route descriptions. Mental maps constitute the final stage of the understanding process and are simulated by graphic sketches. Some problems concerning text-to-image translation of route descriptions are discussed. We present our processing model and the underlying knowledge representation, as well as the relevant cognitive background.

1. INTRODUCTION

Spatial information has received a lot of attention both in theoretical and applied research concerned with the interaction between natural language and images. Descriptions of spatial relations are kinds of texts that can be translated into pictorial mode. Empirical data studied by psychologists suggest analog, imagelike formats for the mental representations of spatial information. This holds for information extracted from perceptual data as well as from linguistic forms (Denis, 1991; Denis & de Vega, 1993;

Tversky, 1993). The terms *mental image* and *mental* or *cognitive map* are often used with respect to this kind of representation (Denis, 1991).

The present work is concerned with route descriptions (RDs) and their translation into graphic sketches, simulating mental representations of the spatial information contained in the text. Our approach is based on two complementary points of view.

The first one can be labeled as a *multimodal approach*. It consists of investigating the relations between the linguistic and the pictorial modes, in particular with respect to information concerning navigation. Graphic sketches and verbal descriptions are two forms frequently used in every-day communication in order to describe a route. In this research, we are interested in exploring the possibilities and constraints in translating linguistic RDs into graphic representations.

The second point of view is the one held by the cognitive scientist. It is concerned with modeling the understanding process, which consists of "translating" a text into its corresponding mental representation. We assume that mental representations take in the case of route descriptions the form of mental maps, in the sense that they reconstruct, in a qualitative way, spatial relations between the elements of the environment.

In the following, we first outline some of the problems that may occur while translating descriptions into pictorial representations. Before presenting our RD-processing model, we discuss the relevant cognitive assumptions concerning mental representation of spatial information and discourse comprehension.

2. TEXT-TO-IMAGE TRANSLATION

Automatic text-image transcription has recently become an active field of study, prompting research and applications concerning these two modes of representation and their possible bidirectional mappings. Different types of texts and images have been considered: narrative texts and motion pictures (Abraham & Desclés, 1992; Kahn, 1979), spatial descriptions and three-dimensional sketches (Arnold & Lebrun, 1992; Yamada, Yamamoto, Ikeda, Nishida, & Doshita, 1992), two-dimensional spatial scenes and linguistic descriptions (André, Bosch, Herzog, & Rist, 1987), and two-dimensional image sequences and linguistic reports (André, Herzog, & Rist, 1988; Schirra & Stopp, 1993). All these texts and images contain spatial information.

There also exist multimodal projects concerning RDs. Their goal is to assist people in navigation. These projects are mainly concerned with the automatic generation of RDs on the basis of visual inputs. The VITRA GUIDE project, for example, deals with the incremental transcription of visual data into texts, such as the ones produced by a codriver (Gapp & Maaß, 1994; Herzog, Maaß, & Wazinski, 1993).

Translation of texts into images is not an easy task as these two modes differ in their expressive capabilities (Arnold, 1990; Arnold & Lebrun, 1992). For example, natural language allows for expression of abstract information, like aspect, tense, modality. It can also convey incomplete or vague information, whereas the pictorial mode requires precise information as, for example, concerning the location of objects. Pictures are better suited to represent complex spatial configurations, allowing one to group information that would occur in a scattered way in a text, due to the linear character of language.

Even when some content can be naturally conveyed by both modes, as in the case of describing a route, there are still differences between the two means of expression, which may lead to informational divergence. This being so, one can expect to encounter problems when trying to encode pictorially the information contained in a verbal description. The following example, taken from the French corpus of RDs (Gryl, 1992) which we used for analysis, illustrates some of the problems that appear when one wants to translate the information contained in a RD into a picture.

> Example 1: *À la sortie des tourniquets du RER tu prends sur ta gauche. Il y a une magnifique descente à prendre. Puis tu tournes à droite, tu tombes sur une série de panneaux d'informations. Tu continues tout droit en longeant les terrains de tennis et tu tombes sur le bâtiment A.* (Translation: "At the turnstiles of the RER station you turn left. You take the steep downgrade. Then you turn right and you come across a series of sign posts. You continue straight on, walking past the tennis courts, and you come to building A.")

The preceding text contains several ambiguities, which may, of course, pose problems for its graphic representation. The most striking case is the relative position of the tennis courts: We do not know on which side of the path they are located (right or left).

Other kinds of ambiguities are due to the fact that in RDs the whole path need not be linguistically expressed. Consider the following two fragments: *tu prends sur ta gauche* ("you turn left"), and *il y une magnifique descente à prendre* ("you take the steep downgrade"). It is not clear whether the downgrade is right after the turn, or "a little farther." The same question holds for the right turn and the signposts, mentioned in the next sentence: Should the posts be represented as immediately following the turning point, or should there be a path between them? This kind of ambiguity is not really perceived unless one wants to derive a graphic representation of the route. The information is complete enough for a real-life situation of finding one's way.

Another kind of problem concerns the *magnifique descente* ("steep down-grade"). It would not be easy to represent a slope in a simple sketch and, even less so its "steepness," which the French word *magnifique* suggests in this context. The incompleteness of information will occur on the graphic side this time, because not all the properties of the described element will be expressed.

Such transcription constraints make the generation of "faithful" graphic representations difficult. It seems that some ambiguity problems can be solved through an analysis of the examples in relation to the physical environments they describe. For example, this applies in cases where we hesitate to decide whether there is a significant stretch of path between two elements of the environment *(landmarks)*, or a turning point and a landmark, mentioned in the text as occurring one after another.

With regard to the indetermination concerning the location of objects, we may choose an arbitrary value, or try to preserve the ambiguity in the graphic mode, which seems quite difficult, though. We should also accept the fact that not all available linguistic information will be expressed in the graphic sketch (e.g., *magnifique descente*).

These are just some of the problems concerning the translation of RDs into maps. There are also types of verbal information that are not at all representable by images, such as comments or evaluations; for example, "you can't miss it" or "it's very simple."

3. COGNITIVE BACKGROUND

According to some cognitive psychologists, spatial information should preferably be encoded in memory in an analog form (Denis, 1991; McNamara, 1986). The popular metaphor for the representations people have of their environments is a mental, or a cognitive map. However, it has been argued that the term *map* is too restrictive. According to Tversky (1993), it is unlikely that the information people acquire about the environment can be organized into a single, maplike cognitive structure. She proposed two other metaphors for spatial representations: *cognitive collage* and *spatial mental model*. The first one applies to the representations that are not coherent, as they contain partial information from different points of view. The second one refers to the more accurate and coherent mental representations of spatial layouts for simple or better known environments. According to the author, the term *spatial mental model* is preferable to *cognitive map*, because the latter suggests that the metric information is preserved.

We use the term *mental map* for mental, imagelike representations generated from natural language route descriptions. We do not imply the reconstruction of metric information, which would not be possible because the source RDs contain qualitative rather than quantitative information. Also, we do not consider absolute coherence in mental maps, such as the one characterizing real maps. It will mostly depend on the level of coherence and determination of the source RDs themselves. Although we are concerned with mental maps, we do not exclude that there may be other modes of representation for the information contained in RDs. Perrig and Kintsch (1985), for example, supposed that RDs are mentally represented as lists of procedure instructions. We will need more psychological evidence concerning the format of people's representations of such descriptions in order to decide whether maplike graphic sketches are, at least to some extent, appropriate to simulate them.

Another aspect of mental representation that we take into account is the dissociation of spatial and visual components of mental imagery (Farah, Hammond, Levine, & Calvanio, 1988). According to psychological as well as neurological evidence, the visual appearance and the spatial location representations are subserved by separate imagery systems. The same distinction has been claimed for visual perception: Two functionally and anatomically independent visual systems have been distinguished (Ungerleider & Mishkin, 1982), namely the "what" system, concerned with the visual appearance of stimuli, and the "where" system, concerned with the spatial layout of stimuli. Farah et al. showed that the debate over whether mental images are visual or spatial is unfounded, because mental imagery has both spatial properties and visual properties. The participation of the visual or the spatial component in a mental representation is dependent on the task being performed.

We make the hypothesis that in the mental images generated from RDs it is the spatial component that is primary. As for the participation of the visual component, it is probably conditioned by the degree of the listener's familiarity with the described environment. However, it is also possible that visual representations come into play even when the environment is totally unknown. Here again, we will need more psychological data concerning the images people actually generate from RDs in order to accurately simulate them. For the time being, however, we choose to represent in mental maps the spatial rather than the visual information available in RDs.

Another area of cognitive science relevant to our problem is discourse comprehension. In order to retrieve information from linguistic RDs and generate mental maps, we base our model on the comprehension principles proposed by psychologists. According to Johnson-Laird (1983) and Van Dijk and Kintsch (1983), people who read or listen to a text build representations on several levels, including a representation of the text itself and a representation of the object or situation described in the text.

This belief is shared by many other cognitive psychologists. In Hegarty and Just (1993), for example, similarly as in Johnson-Laird's theory, these two levels of representation are referred to as the "text-based representation" and the "mental model of the referent." The linguistic representation is considered to be close to the text and the structure of the sentences. The mental model is supposed to be created in working memory by integrating the information coming from the text and elaborating this information using the knowledge activated from the long-term memory. Having distinguished these two levels of representation, the concern of cognitive psychologists is to determine their formats and properties and the way they are elaborated (M.-F. Ehrlich & Tardieu, 1993). The existing hypotheses and assumptions need to be not only confirmed, but also enriched and refined.

In view of the psychological assumptions, we decided to include in our RD-processing model two main levels of representation that we call *textual representation* and *conceptual representation*. The processing consists in the translation of the linguistic description into the conceptual representation, via the textual representation. We use the word *translation* because we consider, following S. Ehrlich (1982), that there is no direct correspondence between the elements of the linguistic expression on one hand, and the conceptual entities on the other hand. Thus, there necessarily is translation in the process of transition from a verbal message to the corresponding conceptual representation. In the final stage of the processing, the spatial information from the conceptual representation is visualized in the *mental map*.

4. ROUTE DESCRIPTION-PROCESSING MODEL

We now describe our model for automatic translation of RDs (input) into mental maps (output), and the underlying knowledge representation. Figure 11.1 shows the main stages of the processing and the knowledge that is used.

4.1. The Textual Representation

The first stage of the RD processing is a *textual representation* (TR), which is the representation of the input description. The automatic analysis of the text yielding the TR has a syntactic and a semantic component. This part of the processing is based on the prototypic linguistic model of the route description, called *route description model* (RDM), which has been elaborated as a result of our analysis of the corpus. According to our linguistic formalization, a RD is composed of two kinds of "global units,"

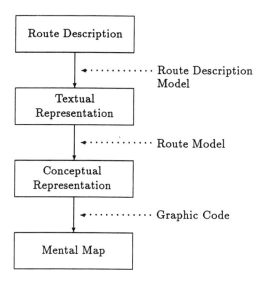

FIG. 11.1. Route description-processing model.

namely the *sequences* and the *connections*. We give next another example of a RD, and we present its division into global units in Fig. 11.2.

Example 2: *Il faut sortir de la gare, prendre la passerelle, descendre la passerelle. Et là, on arrive donc à l'entrée de l'université. Et donc, il faut longer tous les bâtiments, en fait il faut aller toujours tout droit, et on entre par le bâtiment B qui se trouve sur la gauche, où il y a une cafétéria. Là, on va sur la droite et on arrive au bâtiment A.* (Translation: "You have to go out of the station, take the footbridge, and go down the footbridge. Then, you arrive at the entrance of the university. You have to go alongside all of the buildings, in fact, you have to go straight on, and you enter through building B, which is situated on the left, where there is a cafeteria. There, you go to the right and you arrive at building A.")

	connection	sequence
1		*il faut sortir de la gare*
2		*prendre la passerelle*
3		*descendre la passerelle*
4	*et là*	*on arrive donc à l'entrée de l'université*
5	*et donc*	*il faut longer tous les bâtiments*
6	*en fait*	*il faut aller toujours tout droit*
7	*et*	*on entre par le bâtiment B qui se trouve sur la gauche*
8	*où*	*il y a une cafétéria*
9	*là*	*on va sur la droite*
10	*et*	*on arrive au bâtiment A*

FIG. 11.2. Division of a route description into global units.

The sequences are functional-and-thematic units of the RD. Their role may be to *prescribe* actions, to *indicate* objects, to *describe* objects, and so on. A sequence may be realized by an independent clause (e.g., 1: *il faut sortir de la gare*), by a subordinate clause (e.g., 8: *(où) il y a une cafétéria*), or by a sentence consisting of one independent and one subordinate clause when the latter is introduced by a relative pronoun (e.g., *qui, que*) qualifying the noun element of the independent clause (e.g., 7: *on entre par le bâtiment B qui se trouve sur la gauche*).

The function of the connections is to reinforce the structure of the description and to specify the relations between the sequences. A connection may be composed of a *connector* or of a combination of connectors. We classify as connectors different elements that may introduce sequences: conjunctions and conjunctive expressions (e.g., *et, c'est-à-dire, en fait*), adverbs and adverbial expressions (e.g., *là, ensuite, à un moment donné*), prepositions (e.g., *pour*), and so forth.

The TR contains relevant information explicitly mentioned in the text, in the sequences and connections. Our semantic analysis of RDs allowed us to define different kinds of semantic entities, or "local units" of the sequences, which must be represented at this level. Among these entities there are *actions* and *indicators of objects*, which correspond respectively to the verbs expressing prescribed "actions" (e.g., *aller, prendre* la rue, *tourner* à droite: *go, take* a street, *turn* right), and to those introducing "objects" of the environment (e.g., *arriver* à X, *tomber* sur X, *voir* X: *arrive* at X, *come* to X, *see* X). Other semantic entities contained in the sequences are *objects*, expressed by nominal groups, and *relations*, expressed by prepositions.

These local units of RDs are described in the TR in terms of attributes and values. The nouns, for example, indicate the object's attribute *Name*. The objects can also have other attributes, for example, *Number, Size, Color*. The actions are also represented in the TR by means of attributes and values. The attribute *Class* may be filled in by the value *advance* (e.g., *continuer*), *take-way* (e.g., *prendre la rue*), *take-direction* (e.g., *tourner à gauche*), and the like. Some classes of actions can also have the attribute *Type*. For example, the actions belonging to the *Class advance* may be of the following *Types: progress, traverse, mount, descend, enter, exit*.

The *connectors* have been categorized according to the type of relation they specify: *succession* (e.g., *et, puis, après*), *anchorage* (e.g., *ici, à ce moment là, où*), *overlap* (e.g., *en fait, c'est-à-dire*), *alternative* (e.g., *ou, soit*), and so forth. Some connectors may combine (e.g., *et là, là où*). The value of the connection is an important piece of information, as it may trigger inferences concerning the structure of the route at the next level of representation. For example, the connector *en fait* ("in fact"), which according to our categorization introduces the relation *overlap*, indicates that the sequences it connects refer to the same segment of the route. In the following fragment of Example 2:

> *il faut longer tous les bâtiments*
> (you have to go alongside all of the buildings)
>
> *en fait*   *il faut aller toujours tout droit*
> (in fact you have to go straight on)

the actions expressed by the two sequences correspond to one and the same onward move.

The Textual Representation for Example 2 is given in Fig. 11.3.

4.2. The Conceptual Representation

The next step consists of mapping the information from the TR onto the route model (RM). In this way the *conceptual representation* (CR) of the route is created. In fact, the CR has a double character: On the one hand it is a representation of the RD's informational content, and, on the other hand, a representation of some part of the world. In this respect, it is similar to the conception of Kamp's (1981) discourse representation structures (DRS) and to the mental models of Garnham and Oakhill (1993).

The RM (see Fig. 11.4) is a part of the long-term knowledge, specifying the prototypical elements of the route and their organization. It allows one to make inferences and reconstruct implicit information. It may be considered as a conceptual schema for route descriptions, stored in the long-term memory and directing the comprehension of particular RDs. The RM has been elaborated on the basis of the analysis of the corpus. Similar studies have been carried out for other types of discourse, leading to propositions of corresponding prototypical schemas, for example the pioneering analysis of Propp (1968) for Russian folk stories.

The RM specifies all route elements and their possible attributes. Our conceptualization of the *route* first consisted of dividing it into *path* and *landmarks*. For the sake of formalization, the path has been discretized into the elements called *relays* and *transfers*. A transfer is a fragment of the path not "interrupted" by a relay. A relay is a point that initiates a new transfer (e.g., after a change of direction). *Landmarks* are elements of the environment mentioned in the description (e.g., buildings, crossings, streets). According to our model, a *landmark* may be associated with a *transfer* (e.g., "you go past a café"), with a *relay* (e.g., "you come to a café"), with the closing element *goal* (e.g., "you arrive at building A"), or else with another *landmark* (e.g., "the library is behind the football ground") but only when this latter has already been associated with a

```
(route-description
  (sequence
    (action (class . advance) (type . exit))
    (object (name . "gare") (number . 1)))
  (sequence
    (action (class . take-way))
    (object (name . "passerelle") (number . 1)))
  (sequence
    (action (class . advance) (type . descend))
    (object (name . "passerelle") (number . 1)))
  (connection
    (connector (class . succession))
    (connector (class . anchorage)))
  (sequence
    (indicator-of-object (class . reach))
    (object (name . "entrée") (number . 1)
      (relation (type . of) (object (name . "université") (number . 1)))))
  (connection
    (connector (class . succession))
    (connector (class . anchorage)))
  (sequence
    (action (class . advance) (type . go-along))
    (object (name . "batiments") (number . pl)))
  (connection
    (connector (class . overlap)))
  (sequence
    (action (class . advance) (type . progress) (direction . straight)))
  (connection
    (connector (class . succession)))
  (sequence
    (action (class . advance) (type . enter))
    (object (name . "batiment B") (number . 1)
      (relation (type . on-left))))
  (connection
    (connetctor (class . anchorage)))
  (sequence
    (indicator-of-object (class . be))
    (object (name . "cafétéria") (number . 1)))
  (connection
    (connector (class . anchorage)))
  (sequence
    (action (class . take-direction) (direction . right)))
  (connection
    (connector (class . succession)))
  (sequence
    (indicator-of-object (class . reach))
    (object (name . "batiment A") (number . 1))))
```

FIG. 11.3.   The textual representation for Example 2.

relay or a *transfer* (e.g., "you arrive at the football ground"). The *relays* and *transfers* in the RM are placed in sequence, and the pairs (*relay, transfer*), together with their associated *landmarks,* are called *segments.*

The *landmarks* represented in the conceptual representation inherit the attributes and values from the *objects* contained in the textual representation (e.g., *Name, Number, Size*). The values of the other attributes of landmarks are inferred from the TR, mainly from the *actions* and the *relations.* For example, the landmark's attribute *Type*—which can have either of the five values: *simple, frame, way, exit, entrance*—will be filled in with the value *way,* if the *object* in the TR is introduced by an action

Route	::=	`Route:` {*Segment*} `Goal:` *Relay*;
Segment	::=	`Segment:` *Relay*; *Transfer*;
Relay	::=	`Relay:` [`direction=`*Direction*;] {*Landmark*}
Transfer	::=	`Transfer:` [`dimension=`*Dimension*;] {*Landmark*}
Direction	::=	`straight` | `left` | `right`
Dimension	::=	`small` | `big`
Landmark	::=	`type=`*Type*; [`number=`*Number*;] [`disposition=`*Disposition*;] [`name=`*Name*;]
		[`situation=`*Situation*;] [`order=`*Order*;] [`size=`*Size*;] [`color=`*Color*;]
		[`location:`*Location*;] [`Association:`*Association*;]
Type	::=	`simple` | `frame` | `way` | `entrance` | `exit`
Situation	::=	`intra` | `extra` | `bi-extra` | `para-extra` | `infra` | `supra`
Number	::=	`1` | `2` | ...
Disposition	::=	`group` | `series`
Order	::=	*Number* | `last` | `last-but` *Number*
Size	::=	`small` | `big` | *Number* `meters` | *Number* `minutes`
Color	::=	`red` | `blue` | ...
Location	::=	`type:`*Loc-type*; [`Referent:`*Landmark*]
Association	::=	`type:`*Assoc-type*; `Referent:`*Landmark*
Loc-type	::=	`on-left` | `on-right` | `in-front` | `behind` | ...
Assoc-type	::=	`of` | `with` | `without` | `equal`
Name	::=	`"string of characters"`

Notation: typewriter type style - terminal, *italic* type style - nonterminal, { } - repetition, [ ] - optional.

FIG. 11.4. The Route Model.

take-way (e.g., "you take the footbridge"). The *Type frame* is attributed to the landmarks corresponding to the objects associated with an action of the *Type exit, enter, traverse*, that is, the landmarks that constitute a frame for some part of the route. Thus, the *frame* corresponds to the places that one enters and to those that one must traverse (e.g., "you enter the university"; "you go across the football ground"). The attribute *Situation* specifies the general position of the landmark with respect to the path, by means of the following values: *intra*, for the landmarks situated on the path (e.g., "you go across the football ground"); *extra*, for the landmarks situated on a side of the path (e.g., "you pass by a café"); *para-extra*, for the landmarks that are parallel to the path (e.g., "you walk alongside the tennis court"); *biextra*, for the landmarks situated on both sides of the path (e.g., "you pass between two buildings"); *supra*, for the landmarks situated above the path (e.g., "you go under the bridge"); and *infra*, for the landmarks situated beneath the path (e.g., "you go over the river").

In order to be able to build a CR from the TR, we need rules specifying the relationships between the elements constituting these two representations. For example, the class of actions *advance* correspond to a *transfer*, an *object* corresponds to a *landmark*, an action *exit* or *enter* corresponds to a landmark *exit* or *entrance*, respectively. The CR elements are reconstructed on the basis of the TR, but they may also be inferred from the prototypical structure of the route (i.e., from the *route model*). For example, because every route *segment* contains one *relay* slot (which, itself, has the slots for *direction* and *landmarks*) and one *transfer*, we know that there must be a transfer between the place where one turns and the

```
Route:
  Segment:
    Relay: ;
    Transfer: dimension = small
      Landmark: type = exit; situation = intra;
                Association: type = of;
                    Referent: Landmark: type = frame;
                                        number = 1; name = "gare";
  Segment:
    Relay:
      Landmark: type = way; number = 1; name = "passerelle"; situation = intra;
      Transfer: ;
  Segment:
    Relay:
      Landmark: type = entrance; situation = intra;
                Association: type = of;
                    Referent: Landmark: type = frame; number = 1;
                                        name = "université";
      Transfer:
        Landmark: type = simple; disposition = series; name = "batiments";
                  situation = para-extra;
  Segment:
    Relay: direction = left;
    Transfer: dimension = small;
      Landmark: type = entrance; number = 1; name = "batiment B"; situation = intra;
      Landmark: type = simple; number = 1; name = "cafétéria"; situation = extra;
  Segment:
    Relay: direction = right;
    Transfer: ;
  Goal:
    Relay:
      Landmark: type = simple; number = 1; name = "batiment A";
```

FIG. 11.5. The conceptual representation for Example 2.

landmark, described in the last fragment of Example 2: *on va sur la droite et on arrive au bâtiment A* ("you go to the right and you arrive at building A").

The conceptual representation for Example 2 is given in Fig. 11.5.

4.3. The Mental Map

Once the information is integrated and reconstructed within the CR, the mental map (MM) can be generated. During this step, the route reconstructed at the previous level is visualized by means of graphic objects.

The graphic code we are using is simple and symbolic (see Fig. 11.6). More elaborate objects might be introduced in order to allow for a more refined representation of landmarks. For the time being, we have symbols for landmarks that correspond to their *Types* (*simple, frame, way, entrance, exit*). Specific information concerning a landmark, represented by its *Name*, is contained in the textual label that is associated in the sketch with the graphic object. Some of these labels could be replaced by specific icons, at least for the most typical landmarks. So far, however, we have been more interested in recovering the information concerning the landmarks

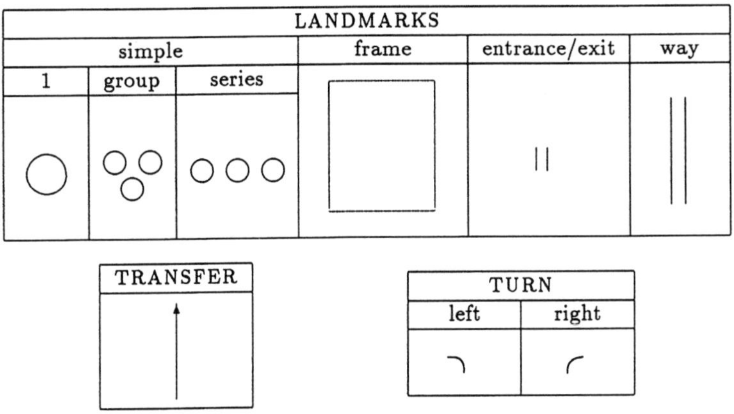

LANDMARKS					
simple			frame	entrance/exit	way
1	group	series			

FIG. 11.6. The graphic objects used for mental maps.

from the context, rather than in the information contained in the nouns that "name" them. Thus, for the time being, many different landmarks are represented by the same *Type*, and consequently by the same graphic symbol. For example, a circle may represent a building as well as a tennis court, in the case when they are both classified as *simple* landmarks (e.g., in: "you pass by a building" and "you pass by a tennis court"). The graphic frame serves to represent a landmark *frame*, that is a landmark that encloses a fragment of the route. A *transfer* is graphically represented by an arrow. A *relay*, which is a point initiating a new transfer, is not represented as such. In the case when a *relay* contains a change of direction, we introduce a graphic object called "turn."

Using these graphic objects, we try to translate the RDs from our corpus into graphic sketches. We encounter, however, the same translation problems as discussed in section 2. Again, in the case of the RD that we considered in this section (cf. Example 2), we realize that an apparently clear and coherent description may contain indetermination, which makes difficult the translation of its content into a graphic representation. Concerning Example 2, these problems are the following. First of all, we do not know how to position the footbridge *(passerelle)* with respect to the exit of the station. We have the same problem concerning the precise location of the buildings *(bâtiments)*: we know that they are situated along the path, but we do not know whether they are on its left or on its right side. The number of buildings is not specified either. There is also a problem concerning the cafeteria's location. We know that it is somewhere near building B, but is it inside or outside, and where exactly? In order to generate graphic maps from such "imperfect" descriptions, we have to decide how to deal with linguistic indetermination. Of course, if we want our processing to be based on cognitive grounds, we need to have

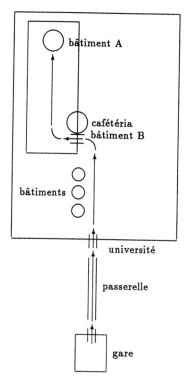

FIG. 11.7. The graphic mental map for Example 2.

an idea of the way people represent such imprecise spatial information, and whether, in case of indetermination, they are able to construct imagelike representations at all.

In Fig. 11.7, we propose a preliminary version of a graphic MM corresponding to the RD of Example 2. The ambiguities have been solved by making arbitrary decisions.

5. CONCLUSION AND PERSPECTIVES

We have proposed a model for automatically translating route descriptions into mental maps, motivating our representation choices with respect to cognitive considerations. We have also discussed some problems related to text-to-image translation. The model has been partially implemented in Gnu Emacs LISP. However, it still needs some adjustments and refining, in particular with respect to the rules coding the relationships between the semantic and the conceptual entities, necessary for the automatic transition from the textual representation to the conceptual representation. Besides, we would like to analyze other corpora of RDs, preferably in other languages than French, in order to validate our prototypical representations. We also have to decide how to deal with the linguistic indetermination of information in order to be able to generate graphic maps from real route descriptions. This final part of the job has been the least developed so far and will be one of the main goals of our efforts in the near future.

ACKNOWLEDGMENTS

I am indebted to Gérard Ligozat for his helpful advice and suggestions. I would also like to thank Wojtek Fraczak for his help with the implementation of a part of the model.

REFERENCES

Abraham, M., & Desclés, J.-P. (1992). Interaction between lexicon and image: Linguistic specifications of animation. In *Proceedings of COLING-92, Nantes, France* (pp. 1043–1047). ICCL.

André, E., Bosch, G., Herzog, G., & Rist, T. (1987). Coping with the intrinsic and the deictic uses of spatial prepositions. In K. Jorrand & L. Sgurev (Eds.), *Artificial intelligence II: Methodology, systems, applications* (pp. 375–382). Amsterdam: North-Holland.

André, E., Herzog, G., & Rist, T. (1988). On the simultaneous interpretation of real world image sequences and their natural language description: The system SOCCER. In *Proceedings of the 8th ECAI* (pp. 449–454). Munich.

Arnold, M. (1990). Transcription automatique verbal-image et vice versa. Contribution à une revue de la question [Automatic verbal-image and vice-versa transcription. Contribution to the review of the problem]. In *Proceedings of EuropIA-90* (pp. 30–37). Paris: Hermès.

Arnold, M., & Lebrun, C. (1992). Utilisation d'une langue pour la création de scènes architecturales en image de synthèse. Expérience et réflexions [Using a language to create architectural scenes in synthesized images. Experiment and reflection.]. *Intellectica, 3*(15), 151–186.

Denis, M. (1991). *Image and cognition*. New York: Harvester-Wheatsheaf.

Denis, M., & de Vega, M. (1993). Modèles mentaux et imagerie mentale [Mental models and mental imagery]. In M.-E. Ehrlich, H. Tardieu, & M. Cavazza (Eds.), *Les modèles mentaux. Approche cognitive des représentations* (pp. 79–100). Paris: Masson.

Ehrlich, M.-F., & Tardieu, H. (1993). Modèles mentaux, modèles de situation et compréhension de textes [Mental models, situation models, and text comprehension]. In M.-E. Ehrlich, H. Tardieu, & M. Cavazza (Eds.), *Les modèles mentaux. Approche cognitive des représentations* (pp. 47–77). Paris: Masson.

Ehrlich, S. (1982). Construction d'une représentation de texte et fonctionnement de la mémoire sémantique [Construction of a text representation and functioning of the semantic memory]. *Bulletin de Psychologie, XXXV*(356), 655–671.

Farah, M., Hammond, K., Levine, D., & Calvanio, R. (1988). Visual and spatial mental imagery: Dissociable systems of representation. *Cognitive Psychology, 20*, 439–462.

Gapp, K.-P., & Maaß, W. (1994). Spatial layout identification and incremental route descriptions. In *Proceedings of the AAAI-94 Workshop on Integration of Natural Language and Vision Processing, Seattle, WA* (pp. 145–152). AAAI Press.

Garnham, A., & Oakhill, J. (1993). Modèles mentaux et compréhension de textes [Mental models and text comprehension]. In M.-E. Ehrlich, H. Tardieu, & M. Cavazza (Eds.), *Les modèles mentaux. Approche cognitive des représentations* (pp. 23–46). Paris: Masson.

Gryl, A. (1992). *Opérations cognitives mises en oeuvre dans la description d'itinéraires* [Cognitive operations used in route description]. Mémoire de DEA, Université Paris 11, France.

Hegarty, M., & Just, M. (1993). Constructing mental models of machines from text and diagrams. *Journal of Memory and Language, 32,* 717–742.

Herzog, G., Maaß, W., & Wazinski, P. (1993). Utilisation du langage naturel et de représentations graphiques pour la description d'itinéraires [Using natural language and graphic representations for route descriptions]. In *Actes du colloque interdisciplinaire du CNRS. Images et langages. Multimodalité et modélisation cognitive* (pp. 243–251). Paris: CNRS.

Johnson-Laird, P. (1983). *Mental models.* Cambridge, MA: Harvard University Press.

Kahn, K. (1979). *Creation of computer animation from story descriptions* (Tech. Rep. No. 540). Cambridge, MA: MIT, Artificial Intelligence Laboratory.

Kamp, H. (1981). A theory of truth and semantic representation. In J. Groenendijk, T. Janssen, & M. Stokof (Eds.), *Formal methods in the study of language* (pp. 277–322). Amsterdam: Mathematical Center Tracts.

McNamara, T. (1986). Mental representations of spatial relations. *Cognitive Psychology, 18,* 87–121.

Perrig, W., & Kintsch, W. (1985). Propositional and situational representations of text. *Journal of Memory and Language, 24,* 503–518.

Propp, V. IA. (1968). *Morphology of the folktale.* Austin, TX: University of Texas Press.

Schirra, J., & Stopp, E. (1993). ANTLIMA—A listener model with mental images. In *Proceedings of IJCAI-93, Chambéry, France* (pp. 175–180).

Tversky, B. (1993). Cognitive maps, cognitive collages and spatial mental models. In *LNCS, 716. Spatial information theory* (pp. 14–24). Berlin: Springer-Verlag.

Ungerleider, L., & Mishkin, M. (1982). Two cortical visual systems. In D. J. Ingle, M. Goodale, & R. Mansfield (Eds.), *Analysis of visual behavior* (pp. 549–586). Cambridge, MA: MIT Press.

Van Dijk, T., & Kintsch, W. (1983). *Strategies of discourse comprehension.* New York: Academic Press.

Yamada, A., Yamamoto, T., Ikeda, H., Nishida, T., & Doshita, S. (1992). Reconstructing spatial image from natural language texts. In *Proceedings of COLING-92, Nantes, France* (pp. 1279–1283). ICCL.

12

Integration of Visuospatial and Linguistic Information: Language Comprehension in Real Time and Real Space

Michael J. Spivey-Knowlton
Cornell University

Michael K. Tanenhaus
University of Rochester

Kathleen M. Eberhard
Notre Dame University

Julie C. Sedivy
University of Rochester

Many psycholinguistic theories postulate that as a spoken linguistic message unfolds over time, it is initially processed by modules that are encapsulated from information provided by other perceptual and cognitive systems. However, we observed immediate effects of relevant *visual* context on the rapid mental processes that accompany spoken language comprehension by recording eye movements using a head-mounted eye-tracking system while subjects followed instructions (containing spatial prepositions) to move real objects around on a table. Under conditions that approximate an ordinary language environment, the visual context influenced spoken word recognition and mediated the resolution of prepositional phrase ambiguity, even during the earliest moments of language processing. These results suggest that approaches toward mapping spatial language onto spatial vision may be most successful with early simultaneous integration of provisional (or probabilistic) interpretations from both modalities.

1. INTRODUCTION

It is often claimed that early stages of language comprehension are comprised of informationally encapsulated modules devoted to processing particular subdomains of the linguistic input without influence from other perceptual and cognitive systems (e.g., Ferreira & Clifton, 1986; Fodor, 1983). In contrast, constraint-based approaches assume that "correlated constraints" from various information sources are immediately integrated during the processing of linguistic input (e.g., MacDonald, Pearlmutter, & Seidenberg, 1994; McClelland, 1987; Spivey-Knowlton & Sedivy, 1995; Tanenhaus & Trueswell, 1995). The temporary ambiguities that arise because language unfolds over time have provided the primary empirical testing ground for evaluating these contrasting theoretical perspectives, with the strongest evidence for information encapsulation coming from studies in which potentially relevant constraints, that are introduced by a prior linguistic context, appear to have delayed effects (cf. Ferreira & Clifton, 1986; Rayner, Garrod, & Perfetti, 1992). However, the "context" in such studies is frequently rather impoverished, consisting of a few sentences that precede the target sentence. In addition, the subject's task is often vague, with no well-defined behavioral goal. Under these conditions, the context may not be perceived as relevant by the comprehender, and even if it is, it must be stored in memory, thus it may not be immediately accessible when the ambiguity is first encountered. Moreover, because the context is introduced linguistically, it is always possible to preserve modularity by expanding the scope of the linguistic module.

The research presented here explores the resolution of temporary ambiguity during the comprehension of spoken language under conditions in which: (a) There is a strong test of modularity because the context comes from a completely different perceptual modality: vision, (b) the situational context is immediately relevant because an action must be carried out that directly relates the utterance to the context, and (c) the context, because it is visual, is *copresent* with the linguistic input, and thus can be interrogated when the ambiguity is first encountered (instead of requiring a memory search). We make use of an experimental paradigm in which the listener follows spoken instructions to manipulate real objects in a display while we record their eye movements—all under conditions that approximate an ordinary language environment (Spivey-Knowlton, Sedivy, Eberhard, & Tanenhaus, 1994; Tanenhaus, Spivey-Knowlton, Eberhard, & Sedivy, 1995). In this experimental paradigm, eye movements to relevant objects are closely time-locked to referential expressions in the spoken instruction. Therefore, this methodology enabled us to observe the effects of visual context on early moments of comprehension, and tap into provisional incremental interpretations that are often observable only in eye-movement patterns (not in hand movements or subjects' intuitions). We anticipate that this methodology will provide useful insights into how humans integrate linguistic and visual information in a wide range of areas. For example, monitoring eye movements may prove useful in understanding people's information-gathering strategies when choosing a frame of reference (Schober, 1995), or when describing relative locations of objects in static scenes (Olivier & Tsujii, 1994), and in dynamic scenes (Gapp & Maaß, 1994). This chapter describes experiments on ambiguity resolution *within individual words* (Experiment 1) and *between alternative prepositional-phrase attachments* (Experiment 2) that reveal early commitments to interpretations based on partial input, and immediate integration of relevant visual information with linguistic information in both word recognition and syntactic parsing.

2. HEAD-MOUNTED EYE TRACKING

Before reaching for an object, people typically move their eyes to fixate it (Ballard, Hayhoe, & Pelz, 1995; Epelboim et al., 1994). Thus, when we instruct a subject to "pick up the candle," she makes a saccadic eye movement to the candle, and our methodology allows us to measure the time elapsed from the beginning of the word *candle* to the initiation of the saccade, as well as record any intermediate fixations to other objects. Eye movements are especially informative about early moments of processing because saccades are relatively automatic and almost entirely ballistic. Thus, an initial misinterpretation of the spoken input, from which the listener rapidly recovers, is still observable as a brief fixation of the "incorrect" object.

Eye movements were monitored by an Applied Science Laboratories (ASL) eyetracker mounted on top of a lightweight helmet. The camera provides an infrared image of the eye sampled at 60 Hz. The center of the pupil and the corneal reflection are tracked to determine the orbit of the eye relative to the head. A scene camera, mounted on the side of the helmet, provides an image of the subject's field of view. Gaze position (indicated by cross hairs) is superimposed over the scene camera image and recorded onto a Hi8 VCR with frame-by-frame playback. Accuracy of the gaze position record is about a degree over a range of $+/- 20°$. The video record was coordinated with the audio record for all data analysis.

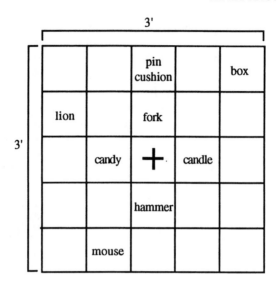

(1)   Look at the cross.

      Pick up the candle.

      Now hold it over the cross.

      Now put it above the mouse.

FIG. 12.1. Example display in which both members of the cohort pair are present in the workspace. (The words in this figure indicate locations of actual objects on the table.)

Subjects were seated at arm's length from a 3 ft-by-3 ft table workspace that was divided into 25 squares (see Fig. 12.1), and were given spoken instructions to move everyday objects around. A black cross in the center square served as a neutral fixation point, where the subject's gaze was directed at the onset of an instruction set. No more than one object was placed in any square, so that noise in the gaze position signal was never enough to mistake a fixation of one object for another. To ensure that subjects did not become aware of the experimental manipulations, the majority of instructions were filler trials that did not involve the experimentally relevant objects.

3. EXPERIMENT 1: AMBIGUITY WITHIN WORDS

3.1. Background. In a classic set of experiments, Marslen-Wilson and colleagues demonstrated that, to a first approximation, recognition of a word occurs shortly after the auditory input uniquely specifies a lexical candidate (for review, see Marslen-Wilson, 1987). For polysyllabic words, this is often prior to the end of the word. For example, the word *elephant* would be recognized shortly after the sound /f/. Prior to that, the auditory input would be consistent with the beginnings of several words, including *elephant, elegant, eloquent,* and *elevator.* Thus, recognition of a spoken word is strongly influenced by the words to which it is phonetically similar, especially those words that share initial phonemes. Marslen-Wilson referred to the set of lexical candidates that is activated in the same phonetic environment as a "cohort."

Evidence from several experimental paradigms indicates that these candidates are partially activated as a word is being processed. For example, cross-modal lexical priming experiments demonstrate that semantic information associated with cohort members is temporarily activated as a word unfolds. The prior context of the utterance and subsequent input provide evidence that is used to evaluate the competing alternatives. Although current models differ in how they account for these data, nearly all models incorporate the idea that the time it takes to recognize a word depends on a set of potential lexical candidates. (See Cutler, 1995, for a recent review.)

This experiment had two goals. The first goal was to determine how closely time-locked eye movements to a target object would be to the name of the object in a spoken instruction. The second goal was to determine whether the presence of a "competitor" object with a similar name would influence eye-movement latencies to the referred-to object. A visually mediated "cohort competitor" effect would provide strong evidence that lexically based information associated with multiple lexical candidates is partially activated during spoken word recognition. In addition, it would demonstrate that relevant visual context affects even the earliest moments of word recognition.

3.2. Procedure. Eight naive subjects participated in this experiment. They were given instructions to pick up an object and then put it in the square above or below another object. A sample instruction set is given in Fig. 12.1.

We used four pairs of objects with names that were phonetically similar until late in the word: *candy/candle, car/carton, penny/pencil,* and *doll/dolphin.* Each critical object appeared with its "cohort competitor" on some trials and with only distractor objects on other trials. Each subject was exposed to two of the four cohort pairs. For instructions involving cohort members, the objects were always in one of the central eight squares (excluding the square with the cross).

Figure 12.1 shows the workspace at the beginning of the instruction set given in Example 1. The instructions and the positions of the objects were varied and counterbalanced to prevent strategies. In particular, we avoided creating any contingencies that would have resulted in predictable instructions.

3.3. Results. Data were analyzed for both the "pick-up" instruction and the "put-it-above" instruction. On all of the critical trials, the movement of the hand to pick up the target object was preceded by a saccade to that object. On 33% of the trials, fixation of the referred-to object was preceded by a saccade launched to an "incorrect" object. For trials without such "false launches," the mean saccade latency for the "pick-up" instruction was 487 ms from the onset of the target word (e.g., *candle*). Saccade latencies were reliably longer when the display contained a "cohort competitor" (530 ms) than when it did not (445 ms); $F(1,7) =$ 9.27, $p < .02$. The average duration of a target word was 300 ms. If we assume that the interval between the onset of programming a saccade and the initiation of the saccade is about 200 ms (Matin, Shao, & Boff, 1993), then we can estimate that the programming of a saccade to the target object began an average of 55 ms before the end of the word in the competitor-absent condition.

More false launches were made when a competitor was present than when it was absent, but this difference was not reliable (37% compared to 29%). Of the false launches in the competitor-present condition, 61% were to the competitor object. In contrast, in the competitor-absent condition, only 25% of the false launches were to the object that occupied the same square as the competitor object in its corresponding competitor-present display. This difference was reliable in an analysis for six subjects; $F(1,5) = 14.90$, $p < .02$. (Two subjects did not make any false launches in the competitor-absent condition and thus were excluded from the analysis.)

The subsequent instruction, to put the object-in-hand into a square above another object (see Example 1), provided an additional test of this visually mediated cohort effect *when the critical word is the object of a spatial preposition*. Similar, though less robust, effects were observed. For example, when the subject was instructed to put a distractor object above a cohort object with its competitor *absent* (i.e., "Now put it [fork in hand] above the candy."), mean saccade latency was 370 ms with 22% false launches. In contrast, when the cohort's competitor was *present* (i.e., candle also on the table), mean saccade latency was 450 ms with 37% false launches. With relatively few data points for the subsequent instruction so far, these differences are only marginally significant. Most interesting, however, is what happens when the subject has a cohort object in hand and is instructed to put it above its competitor (i.e., "Now put it [candle in hand] above the candy."). The results of this condition patterned with the competitor present condition, with a mean saccade latency of 470 ms and 38% false launches. This indicates that subjects were not strategically anticipating cohort comparisons. It also suggests that relatively complex real-world knowledge, such as the fact that an object cannot be placed *above itself*, does not override the visual input's immediate influence on word recognition.

3.4. Discussion. Three critical results emerged from this experiment. First, eye movements to the target object were closely time-locked to the linguistic expression that referred to that object. Thus, the eye movements provide an informative measure of ongoing comprehension. Second, the latency with which the saccades to the target object were launched provides clear evidence that activation of lexical representations begins before the end of a word. The high rate of false launches to competitors lends further support to the idea that multiple lexical candidates are activated early on in recognition. Third, the presence of possible referents in the visual context clearly influenced the speed with which a referent in the speech stream was identified. This demonstrates that the instruction was interpreted incrementally, taking into account the set of relevant objects present in the visual workspace. Thus, in contrast to modular theories of language comprehension, it appears that on-line word recognition is immediately informed by visual perception.

4. EXPERIMENT 2: AMBIGUITY BETWEEN PREPOSITIONAL-PHRASE ATTACHMENTS

4.1. Background. Clearly, the place where the modularity hypothesis has found the most purchase is in the realm of syntactic processing (cf. Ferreira & Clifton, 1986). The strongest evidence for the modularity of syntactic processing has come from studies using sentences with temporary syntactic ambiguities in which readers have clear preferences for particular interpretations that persist momentarily *even when preceding linguistic context supports the alternative interpretation* (e.g., Britt, 1994; Ferreira & Clifton, 1986). In this type of experiment, the context is typically comprised of a few sentences preceding the target sentence. However, when the context is a visual display that is immediately relevant to the linguistic input (because an action is expected), and storing the context in memory is unnecessary (because the visual context is copresent with the spoken input), syntactic processing may indeed show immediate effects of context. If so, this would provide definitive evidence against the encapsulation of syntactic processing.

4.2. Procedure. We used instructions containing the temporary syntactic ambiguity with perhaps the strongest syntactic preference in English, as illustrated by Example 2:

2a. Put the saltshaker on the envelope in the bowl.
2b. Put the saltshaker that's on the envelope in the bowl.

In sentence 2(a), the first prepositional phrase (PP), "on the envelope," is ambiguous as to whether it modifies the noun phrase (NP; "the saltshaker"), thus specifying the Location of the object to be picked up, or whether it denotes the Goal of the event, that is, where the saltshaker is to be put. Readers' and listeners' preferred initial interpretation of the first PP is that it specifies the Goal, resulting in momentary confusion if the rest of the sentence is inconsistent with that interpretation (cf. Ferreira & Clifton, 1986). In Example 2(b) the word *that's* disambiguates the phrase as a modifier, serving as an unambiguous control condition.

Six naive subjects were presented with six instances of each type of instruction (ambiguous and unambiguous) illustrated in Example 2, with a *one-referent* visual context that supported the Goal interpretation or a *two-referent* context that supported the Location-based modification interpretation. In the *one-referent* context for this example, the workspace contained a saltshaker on an envelope, another envelope, a bowl, and an apple. Upon hearing the phrase "the saltshaker," subjects can immediately identify the object to be moved because there is only one saltshaker and thus they are likely to assume that "on the envelope" is specifying the Goal of the putting event. In the *two-referent* context, however, the apple was replaced by a second saltshaker that was on a napkin. Thus, "the saltshaker," could refer to either of the two saltshakers and the phrase "on the envelope" provides modifying information that specifies which saltshaker is the correct referent.

This particular type of context has been characterized in terms of referential presuppositions that are made by definite noun phrases, such as "the saltshaker" (Altmann & Steedman, 1988; Crain & Steedman, 1985). For example, "the saltshaker," without any postmodifying phrase, presupposes a unique saltshaker in the context. However, "the saltshaker on the envelope," presupposes a *set of saltshakers* in context, one of which is unique by virtue of its being "on an envelope." According to Steedman and colleagues, it is via the contextual satisfaction of these presuppositions that referential processing determines the resolution of the syntactic ambiguity (in our case, by directly linking the utterance to the visual display). If this kind of context, provided in a visual display, has immediate effects on the processing of syntactic ambiguity, then the notion of an encapsulated language processing system (especially an encapsulated syntactic processor) is severely compromised.

4.3. Results. Strikingly different fixation patterns between the two visual contexts revealed that the ambiguous phrase "on the envelope" was initially interpreted as a Goal in the one-referent context, but as a modifier in the two-referent context. In the one-referent context, subjects looked at the incorrect Goal (e.g., the irrelevant envelope) on 55% of the trials shortly after hearing the ambiguous PP, whereas they never looked at it during the unambiguous instruction; $t(5) = 4.11$, $p < .01$. In contrast, when the context contained two possible referents, subjects rarely looked at the incorrect Goal (17% of the trials), and there was no difference between the ambiguous and unambiguous instructions. The statistical interaction between Context and Ambiguity was reliable; $F(1,5) = 8.24$, $p < .05$.

Figures 12.2 and 12.3 summarize the most typical sequences of eye movements and their timing in relation to words in the ambiguous and unambiguous instructions for the one-referent and the two-referent contexts. In the one-referent context with the ambiguous instruction, subjects first looked at the target object (the saltshaker) 500 ms after hearing *saltshaker*, then looked at the incorrect Goal (the upper-right envelope) 484 ms after hearing *envelope*. In contrast, with the unambiguous instruction, the first look to a Goal did not occur until 471 ms after the subject heard the word *bowl*. Example 3 demonstrates the approximate temporal

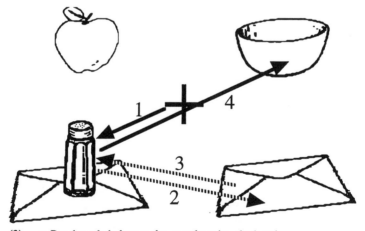

(3)   a. Put the saltshaker on the₁ envelope in₂ the bowl.₃   ₄

     b. Put the saltshaker that's ₁on the envelope in the bowl.₄

FIG. 12.2. Typical sequence of eye movements in the one-referent context for the ambiguous and unambiguous instructions. Dashed arrows show the intermediate saccades to the incorrect Goal and back to the referent object that occur *only in the ambiguous instruction*. (See Examples 3a and 3b for the temporal relationship between eye movements and words in the speech stream.)

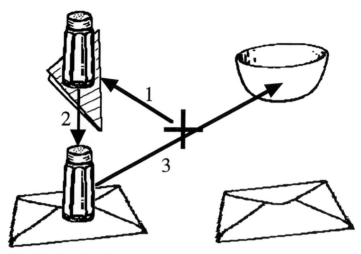

(4) a. Put the saltshaker on the $_1$envelope in $_2$the bowl.$_3$

b. Put the saltshaker that's on$_1$ the envelope$_2$ in the bowl.$_3$

FIG. 12.3. Typical sequence of eye movements in the two-referent context. Note that, in this context, the eye-movement pattern did not differ for the ambiguous and unambiguous instructions. (See subscripted indices in Examples 4a and 4b for the temporal relationship between eye movements and words in the speech stream.)

relationship between eye movements and words in the speech stream, as indicated by subscripted indices to eye movements shown in Fig. 12.2.

In the two-referent context, subjects often looked at both saltshakers, reflecting the fact that the referent of "the saltshaker" was temporarily ambiguous. Subjects looked at the incorrect object (other saltshaker) on 42% of the unambiguous trials and on 61% of the ambiguous trials. In contrast, in the one-referent context, subjects rarely looked at the incorrect object (apple): 0% and 6% of the trials for the ambiguous and unambiguous instructions, respectively. The three-way interaction among Context, Ambiguity, and Type of Incorrect Eye Movement (object vs. Goal) revealed the bias toward a Goal interpretation in the one-referent context and toward a Location-based modification interpretation in the two-referent context; $F(1,5) = 18.41$, $p < .01$.

In the two-referent context (Fig. 12.3), the timing of eye movements relative to the speech stream was nearly identical for ambiguous and unambiguous instructions. This indicates that subjects were interpreting the PP ("on the envelope") as an NP-modifier (instead of as a Goal) equally quickly in both ambiguous and unambiguous instructions.

In addition to examining the effects of one- versus two-referent contexts on syntactic processing, we examined the effects of a "three-and-one" referent context; that is, instead of having *one* additional referent, there were *three* additional referents. So, in place of the saltshaker on a napkin in Fig. 12.3, there were *three saltshakers* in that square. We did this to examine whether the presupposition of uniqueness associated with the definite determiner *the* (cf. Heim, 1982) in "Put the saltshaker" would bias the subject toward the lone saltshaker (on the envelope). Indeed, such a bias was observed in saccade latencies to the target saltshaker, which resembled those of the one-referent context. Subjects rarely looked at the three saltshakers. However, despite the fact that this context made the referent relatively unique early on in the instruction, it was still plausible given the context (and in fact preferred) to interpret the ambiguous PP "on the envelope" as as a modifier instead of a Goal. This is seen in the latter half of the eye-movement pattern, which resembled the two-referent condition: Subjects rarely looked at the incorrect Goal. Thus, this *3-and-1 referent* context elicited an overall eye-movement pattern similar to that for the *unambiguous* instruction in the one-referent context. This result suggests that the syntactic decision to modify the NP "the saltshaker" is not purely determined by referential non-uniqueness of the NP (e.g., Altmann & Steedman, 1988), but general plausibility of modification also plays a role in initial syntactic processing.

4.4. Discussion.   It is clear from these results that the relevant aspects of the visual scene influenced even the initial moments of syntactic analysis. When an object that is referred to in the speech stream is unique in the visual input, further specification of it is deemed unnecessary, resulting in a bias toward interpreting an ambiguous PP as describing the event and not the object. In contrast, the visual presence of multiple referents (e.g., two or more saltshakers) biases the listener toward interpreting the ambiguous PP as describing *which object* is being referred to, instead of *where to put it*. Crucially, this effect of visual context is observed at the earliest measurable point in processing.

5. GENERAL DISCUSSION

The results of Experiments 1 and 2 demonstrate that, in natural contexts, people seek to establish reference with respect to their intended actions during the earliest moments of linguistic processing. Moreover, referentially relevant nonlinguistic information immediately affects how the linguistic input is initially structured. Given these results, approaches to language comprehension that assign a central role to encapsulated linguistic subsystems are unlikely to prove fruitful. More promising are theories in which grammatical constraints are integrated into processing systems that coordinate linguistic and nonlinguistic information as the linguistic input is processed (e.g., MacDonald et al., 1994; McClelland, 1987; Spivey-Knowlton & Sedivy, 1995; Tanenhaus & Trueswell, 1995). In this view, even the earliest computations on language input are richly contextualized with respect to accompanying actions and relevant entities in the environment (cf. Clark, 1992). These results are especially relevant for the growing interest in computational approaches to integrating language and vision (cf. McKevitt, 1994).

In terms of the representation and processing of spatial expressions, this work demonstrates the fluidity with which people integrate spatial information from the visual and linguistic modalities. For example, when a spatial preposition is ambiguous between describing a *movement location* and a *reference location* (Put the saltshaker on . . .), the visuospatial context can immediately determine the resolution of that ambiguity. In order to incorporate such smooth integration, computational models will need to employ incremental partial interpretations of the linguistic input which, when involving spatial expressions, are simultaneously integrated with a generalized visual representation of the spatial relationship and with the immediate visuospatial context.

The overall findings with this methodology show that, with well-defined tasks, eye movements can be used to observe some of the rapid mental processes that underlie spoken language comprehension during goal-directed action in natural contexts. This paradigm can naturally be extended to explore issues ranging from fine details of spoken word recognition to rich conversational interactions during cooperative problem solving. For example, knowing where speakers look when they are choosing between possible spatial perspectives to take, or between possible relative spatial descriptors to use, may provide some insight into the moment-by-moment cognitive processes involved in mapping spatial language onto spatial vision.

ACKNOWLEDGMENTS

Thanks to Dana Ballard and Mary Hayhoe for encouraging us to use their laboratory, to Jeff Pelz for teaching us how to use the equipment, and to Kenzu Kobashi for assistance in data collection. This work has been supported by NIH resource Grant 1-P41-RR09283, an NSF Graduate Fellowship to Michael Spivey-Knowlton, NIH Grant HD27206 to Michael K. Tanenhaus, and a Canadian SSHRC fellowship to Julie C. Sedivy. Portions of this work were reported previously in *Proceedings of the 17th Annual Conference of the Cognitive Science Society* (1995), in J. McClelland & T. Inui

(Eds.), *Attention & Performance XVI* (1996), and in *Science, 268*, 1632–1634 (1995).

REFERENCES

Altmann, G., & Steedman, M. (1988). Interaction with context during human sentence processing. *Cognition, 30*, 191–238.

Ballard, D., Hayhoe, M., & Pelz, J. (1995). Memory representations in natural tasks. *Journal of Cognitive Neuroscience, 7*, 68–82.

Britt, M. A. (1994). The interaction of referential ambiguity and argument structure in the parsing of prepositional phrases. *Journal of Memory and Language, 33*, 251–283.

Clark, H. (1992). *Arenas of language use.* Chicago: University of Chicago Press.

Crain, S., & Steedman, M. (1985). On not being led up the garden path. In D. Dowty, L. Kartunnen, & A. Zwicky (Eds.), *Natural language parsing* (pp. 320–358). Cambridge, MA: Cambridge University Press.

Cutler, A. (1995). Spoken word recognition and production. In J. Miller & P. Eimas (Eds.), *Handbook of cognition and perception* (pp. 97–136). New York: Academic Press.

Epelboim, J., Collewijn, H., Kowler, E., Erkelens, C., Edwards, M., Pizlo, Z., & Steinman, R. (1994). Natural oculomotor performance in looking and tapping tasks. In *Proceedings of the 16th Annual Conference of the Cognitive Science Society* (pp. 272–277). Hillsdale, NJ: Lawrence Erlbaum Associates.

Ferreira, F., & Clifton, C. (1986). The independence of syntactic processing. *Journal of Memory and Language, 25*, 348–368.

Fodor, J. A. (1983). *Modularity of mind.* Cambridge, MA: MIT Press.

Gapp, K., & Maaß, W. (1994). Spatial layout identification and incremental descriptions. In *AAAI-94 Workshop Proceedings on the Integration of Natural Language and Vision Processing* (pp. 145–152). Seattle, WA: American Association for Artificial Intelligence.

Heim, I. (1982). *The semantics of definite and indefinite noun phrases.* Amherst, MA: GLSA.

MacDonald, M., Pearlmutter, N., & Seidenberg, M. (1994). The lexical nature of syntactic ambiguity resolution. *Psychological Review, 101*, 676–703.

Marslen-Wilson, W. (1987). Functional parallelism in word recognition. *Cognition, 25*, 71–102.

Matin, E., Shao, K. C., & Boff, K. R. (1993). Saccadic overhead: Information processing time with and without saccades. *Perception & Psychophysics, 53*, 372–380.

McClelland, J. (1987). The case for interactionism in sentence processing. In M. Coltheart (Ed.), *Attention & performance* (Vol. 12, pp. 1–36). Hove, England: Lawrence Erlbaum Associates.

McKevitt, P. (Ed.). (1994). Integration of language and vision processing [Special issue]. *Artificial Intelligence Review, 8*(1–3).

Olivier, P., & Tsujii, J. (1994). Prepositional semantics in the WIP system. In *AAAI-94 Workshop Proceedings on the Integration of Natural Language and Vision Processing* (pp. 139–144). Seattle, WA: American Association for Artificial Intelligence.

Rayner, K., Garrod, S., & Perfetti, C. (1992). Discourse influences during parsing are delayed. *Cognition, 45*, 109–139.

Schober, M. (1995). Speakers, addressees, and frames of reference: Whose effort is minimzed in conversations about locations? *Discourse Processes, 20*, 219–247.

Spivey-Knowlton, M., & Sedivy, J. (1995). Resolving attachment ambiguities with multiple constraints. *Cognition, 55*, 227–267.

Spivey-Knowlton, M., Sedivy, J., Eberhard, K., & Tanenhaus, M. (1994). Psycholinguistic study of the interaction between language and vision. In *AAAI-94 Workshop Proceedings on the Integration of Natural Language and Vision Processing* (pp. 189–192). Seattle, WA: American Association for Artificial Intelligence.

Tanenhaus, M., Spivey-Knowlton, M., Eberhard, K., & Sedivy, J. (1995). Integration of visual and linguistic information during spoken language comprehension. *Science, 268,* 1632–1634.

Tanenhaus, M., & Trueswell, J. (1995). Sentence comprehension. In J. Miller & P. Eimas (Eds.), *Handbook of cognition and perception* (pp. 217–262). New York: Academic Press.

13

Human Spatial Concepts Reflect Regularities of the Physical World and Human Body

David J. Bryant
NYMA, Inc., Egg Harbor Township, New Jersey[1]

Our world has a definite geometric structure, but it also has the crucial functional property of a fixed gravitational axis. Our bodies also have a definite structure, with two asymmetric body axes, one of which (our head/feet axis) is normally aligned with gravity. Not surprisingly, people conceptualize space in terms of these three axes and their relation to gravity. The nature of our bodies, with their three axes, and our perceptual and motor apparati, eyes and limbs facing forward, ultimately determine how we define spatial relations. In the strongest sense, our experience in space determines the ways in which it is fundamentally *possible* for us to conceive of space. To a sea urchin, with no bilateral plane of asymmetry, the concepts of front, back, left, and right—so crucial to our way of thinking of space—would have no meaning. Thus, spatial cognition is based not only on the geometry of space, but also on the functional design of our body and its relation to the world. One goal of this chapter is to describe how regularities of the physical world and our perceptual experience have played a fundamental role in determining how we conceive of space.

Our concepts of space carry the stamp of our experience. Nevertheless, we learn about our environment in various ways—through navigation, language, models, diagrams, and so on. Further, we make judgments and actions under various conditions—notably during perception of a place

---

[1]This research was conducted at Northeastern University.

and from memory. These factors too should guide spatial cognition. A second goal of this chapter is to describe how we use spatial concepts in different circumstances.

1. MEMORY AND PERCEPTION
OF A SIMPLE SPATIAL SITUATION

To approach these broad issues, my colleagues Barbara Tversky, Maggie Lanca, and Nancy Franklin, and I have examined one situation in detail. We focused on a prototypic spatial situation, that of a person surrounded by objects, and used an experimental paradigm developed by Franklin and Tversky (1990) and extended by Bryant, Tversky, and Franklin (1992). This was first studied for the self in an array of objects. In the work described here, we considered one person (the subject) looking at another person who was surrounded by objects to their six body sides.

A schematic diagram of this situation is shown in Fig. 13.1. The person is in a natural scene, such as a construction site, standing on a table or stepladder. Six objects are located around the person, all directly aligned with a major body side. Although simple, this situation has ecological validity as well as being tractable. Most of the time, people find themselves in environments with objects located more or less to the sides of their bodies. The scene can be considered from an internal perspective of the person. Alternatively, one could take the perspective of someone outside the scene (as you are doing when viewing the picture).

After subjects studied such a scene, it was further described or depicted with a change—the person in it rotated and/or changed posture. Thus, the person faced different objects and stood upright or reclined. Subjects were presented with direction probes—terms referring to the person's six body sides (front, back, head, feet, left, and right). Subjects named the object currently at the probed direction relative to the person as quickly and accurately as possible. After all directions had been probed, the person would be reoriented and the subject given direction probes for the new orientation. Subjects made few errors, so the data of interest were response times to the six directions. Because certain body axes have a favored status in our interactions with the world, they are more salient to thinking about space. These differences lead to differences in retrieval times and indicate the spatial concepts organizing memory or perception.

This paradigm has been applied to different modes of learning scenes—reading narrative descriptions, viewing a model, and studying a diagram. The question in each case is: What spatial concepts are relevant to memory and perception of the scene? Certainly, spatial cognition need not be the same in both cases—the goals, cognitive mechanisms, and sources of

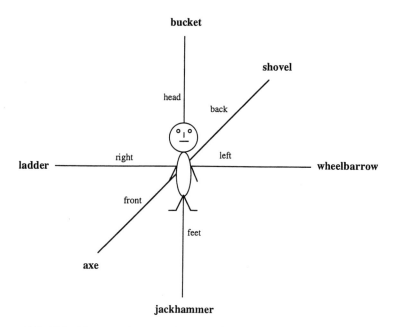

FIG. 13.1. Diagram of prototypic scene. A person is surrounded by six objects (this scene represents a construction site). In some studies, subjects themselves were the person in the scene, whereas in others subjects were external viewers looking at a person in the scene. In narratives, the locations of objects would be described and readers would never see a picture. In perceptual studies, subjects would view a diagram or model depicting the scene.

information are different for memory and perception. We have derived two alternative models of how people conceptualize our prototypic scene. Results so far have indicated that different concepts govern memory and perception of space under different conditions.

1.1. Spatial Framework Analysis

Franklin and Tversky (1990) originally devised the spatial framework analysis to describe the mental models readers derive from narratives. The spatial framework analysis extends analyses by Clark (1973), Levelt (1984), Miller and Johnson-Laird (1976), and Shepard and Hurwitz (1984), among others. According to this theory, subjects construct a spatial framework consisting of mental extensions of the three body axes, head/feet, front/back, and left/right, and associate objects to that framework. The accessibility of an axis depends on characteristics of the body and the perceptual world. For an upright observer, the head/feet axis is most accessible because it is physically asymmetric and correlated with the fixed environmental axis of gravity. The front/back axis is next most accessible. It is not associated with a fixed environmental axis but is strongly asymmetric, separating the world that can be seen and manipulated from the world that cannot be easily perceived or manipulated. The left/right axis is least accessible because it has no salient asymmetries. For the upright observer, people should be fastest to identify objects at the head and feet, followed by front and back, followed by left and right.

The situation changes for a reclining person. The head/feet axis is no longer correlated with gravity, so the accessibility of axes depends solely on their asymmetries. The perceptual and behavioral asymmetries of the front/back axis are stronger than those of the head/feet axis. Thus, for a reclining person, identification along front/back should be faster than head/feet, which should be faster than left/right.

The spatial framework analysis naturally applies to a person in a scene and to second-person narratives describing "you" in a scene. It can, however, also apply to third-person narratives describing scenes in relation to another person. Although a reader could adopt a viewpoint outside the character, this would entail keeping track of two perspectives (your own and the character's). Instead, subjects could adopt the perspective of the character and construct spatial frameworks to represent scenes. In this case, response times to identify objects around the character would conform to the spatial framework pattern because the subject "mentally" occupies the position of the person in the scene. This reduces the reader's mental load by allowing him or her to maintain the single most relevant perspective.

A further extension of the spatial framework analysis is to memory of physical scenes actually surrounding a person. If the same spatial concepts are involved in representing described and observed scenes, we would expect subjects to be faster to head/feet, followed by front/back, followed by left/right for the upright posture, but faster to front/back than head/feet for the reclining posture. In another case, subjects might view a three-dimensional (3-D) model depicting a person in a scene then respond to direction probes by indicating the locations of objects relative to the *other person's* point of view. This situation is analogous to a third-person narrative, and subjects could adopt the perspective of the person in the model. If so, they would exhibit the pattern of response times predicted by the spatial framework analysis.

1.2. Intrinsic Computation Analysis

The spatial framework analysis embodies many important features of the environment and human body. Thus, one might believe that it would account for behavior in all domains, including perception. There is, however, a profound distinction to be made between the processes of memory and perception. In perception, an observer views a person and must locate objects relative to that person. Mentally adopting that person's perspective and creating a spatial framework is only one strategy for solving this task.

According to the intrinsic computation analysis, observers view a person in a scene *intrinsically* (see Levelt, 1984; Miller & Johnson-Laird, 1976). That is, they identify the intrinsic sides of the person by using the same general perceptual mechanisms used in object recognition. Object recognition involves extracting the axes of the object because identification depends on how features are spatially related to one another. Some intrinsic axes are more readily determined than others. Many researchers have demonstrated that the top/bottom axis (the head/feet in humans) is primary in object perception (e.g., Jolicoeur, 1985; Maki, 1986; Rock, 1973). People are faster to identify the top/bottom (head/feet) than the front/back (Jolicoeur, Ingleton, Bartram, & Booth, 1993) and the left/right (Corballis & Cullen, 1986) of objects at all orientations (including reclining). The left/right axis is derived from knowing the top and front sides of an object and is necessarily slowest.

To locate an object, subjects extract directions by analyzing the person's intrinsic axes, then visually scanning in the appropriate direction. On this basis, the main prediction of the intrinsic computation analysis is that an observer will *always* be fastest to identify objects at the head/feet, then the front/back, and finally the left/right of a viewed person, regardless of the person's posture. Thus, the main way to distinguish the use of spatial frameworks from intrinsic computation is to compare patterns of response times for head/feet and front/back across upright and reclining orientations.

What would be the advantage of guiding perception of scenes by the spatial concepts embodied in the intrinsic computation analysis? Intrinsic computation makes use of general perceptual processes specialized for exactly the task we have examined—determining directions within an object-centered frame of reference. Also, it allows an observer to identify directions without mentally placing themselves in another person's perspective or creating a mental spatial framework. This eliminates any conflict between the subject's actual viewpoint and that of the other person.

2. EMPIRICAL RESEARCH

The spatial framework and intrinsic computation analyses are alternative hypotheses about how people's spatial concepts organize and influence behavior. The question is which hypothesis provides the better account of spatial cognition in the various situations and tasks with which people are confronted. In this section, I review empirical research by myself and colleagues concerning narrative comprehension, memory and perception of observed scenes, and perception and memory of diagrammed scenes. As becomes evident, both the spatial framework and intrinsic computation analyses provide good models for behavior in different situations.

2.1. Understanding Narratives

Initially, Franklin and Tversky (1990) sought to characterize spatial representations acquired from texts. In a series of five experiments, subjects read narratives written in the second person describing a simple environment around themselves. For example, subjects read a story that described "you" standing on a balcony in a space museum, with a large map of the solar system hanging on a wall directly in front of you, a spacesuit hanging directly to your right side, a meteorite directly beyond your feet on the first floor of the museum, and so on. After reading an initial description, subjects continued reading the narrative sentence-by-sentence on a computer screen. The text periodically reoriented the subject so that he or she faced different objects, and changed posture, lying down or standing up. Throughout the narrative, subjects were probed with direction terms and responded with the object *currently* located at that direction.

The results of Franklin and Tversky's (1990) experiments were entirely in line with predictions of the spatial framework analysis. In all cases, when upright in the narrative, subjects were faster to head/feet than front/back than left/right, and faster to front than back. When reclining, subjects were faster to front/back than head/feet than left/right. Thus, readers created mental models of the described scenes and these models reflected the enduring physical and bodily regularities that determine human experience with space. These findings served as the first indication that behavior, in this case comprehension and memory for narratives, is strongly influenced by spatial properties of the world and human body.

This is not just true of narratives that specifically single out "you" the reader as the focus of the spatial scene. The spatial framework analysis generalizes to situations in which a reader can mentally adopt the perspective of a person (Bryant et al., 1992). The space museum scene can be described in the third person. Subjects read that Sue is standing on a balcony in a space museum, with a large map of the solar system hanging on a wall directly in front of her, a spacesuit hanging directly to her right side, a meteorite directly beyond her feet resting on the first floor of the museum, and so on. As before, subjects read an initial description then continued reading the narrative sentence-by-sentence. The text periodically reoriented the character so that he or she faced different objects, and changed posture. Throughout the narrative, subjects were probed

TABLE 13.1
Mean Response Times (in Seconds) for Narratives
With Unspecified Perspective and a Central Person

Posture	Direction					
	Head	*Feet*	*Front*	*Back*	*Left*	*Right*
Upright	1.59	1.41	1.53	1.69	2.27	2.40
Pairwise means		1.50		1.61		2.34
Reclining	2.22	2.05	1.67	1.87	3.07	2.46
Pairwise means		2.14		1.77		2.76

Note. From Bryant, Tversky, and Franklin (1992), Experiment 2. Copyright © 1992 by Academic Press. Adapted by permission.

with direction terms that referred to the sides of the *character*. Although subjects could adopt a perspective outside the scene, we expected they would adopt the character's perspective. This simplifies the reader's mental world by allowing them to maintain only one perspective.

The pattern of subjects' response times to direction probes conformed to predictions of the spatial framework analysis, as shown in Table 13.1. Although the narratives allowed multiple perspectives, readers spontaneously brought to bear the ingrained spatial concepts of the spatial framework analysis to create mental models of the scenes. The general result, that, for the upright posture, subjects are faster to head/feet than front/back than left/right, and to front than back, has been replicated when directions are probed for with object names (see Bryant & Tversky, 1992), and for narratives with a central inanimate object, a finding that implies readers can even mentally impose their spatial concepts on non-human entities (Bryant et al., 1992, Experiment 3).

2.2. Observed Scenes

Narrative comprehension is only one way of expressing spatial knowledge, and we became interested in the relation between learning scenes from narratives and from visual observation. In reading a narrative, one must entirely construct the spatial configuration. This may place a strong impetus on the reader to employ spatial frameworks. When observing a scene, subjects do not construct a configuration, and other factors may influence understanding of the spatial relations.

Thus, we (Bryant, Tversky, & Lanca, 1997) posed the question of whether mental models established from narrative are equivalent to mental models established from observation. To answer it, we had subjects learn spatial arrays around themselves or another person from observation. We compared the pattern of response times for responding from memory to the pattern obtained previously for learning scenes from narratives. A second question was whether spatial mental models are like internalized perception. To answer it, we compared the patterns of response times for responding from memory to the pattern obtained when responding from perception. In the first experiment, subjects learned a physical spatial array of objects. Subjects stood or reclined on a bench in an empty room. Large pictures of objects were hung on the walls, ceiling, and floor at the six directions from the subject's body. After learning the scene, subjects responded to direction probes either from memory or while looking at the scene.

When responding from memory, subjects produced the pattern of response times associated with the spatial framework analysis (the pattern observed for narrative comprehension), as illustrated in Table 13.2. This result supports the conclusion that spatial mental models constructed from experience are equivalent to those constructed from descriptions.

A different pattern emerged in the perception condition. Here, subjects had the opportunity to look in probed directions, which they often did. Subjects, however, also responded without looking on numerous trials. Thus, two patterns of data were observed. When subjects did not turn to look when responding, their response times conformed to the spatial framework analysis and reflects memory for the scene. When subjects actually did look at the probed direction to find the object, response times exhibited a *physical transformation* pattern. Specifically, response times to front were fastest, slowest to back, and in between for the other four directions, all 90° from front. The perception condition offered the first indication that the spatial framework analysis is not a universal description of the way people deal with space around themselves.

In a second experiment, subjects viewed a 3-D model of a scene containing a doll surrounded by drawings of objects beyond the doll's head, feet, front, back, left, and right. One group participated in a memory

TABLE 13.2
Mean Response Times (in Seconds) in Memory for Observed Scenes

	Direction					
Posture	*Head*	*Feet*	*Front*	*Back*	*Left*	*Right*
Upright	1.14	1.14	1.18	1.29	1.48	1.46
Pairwise means		1.14		1.24		1.47
Reclining	1.32	1.35	1.26	1.31	1.52	1.46
Pairwise means		1.34		1.28		1.49

Note. From Bryant, Tversky, and Lanca (1997), Experiment 1. Copyright © 1997 by D. Bryant. Adapted by permission.

condition. After subjects studied the model, it was removed from view and subjects responded to direction probes from the doll's perspective from memory. During the procedure, subjects were periodically told that the doll had rotated to face a new object, or had changed posture. Thus, subjects needed to update the current positions of objects relative to the doll to respond to direction probes. We expected that subjects would construct spatial frameworks from the doll's point of view to keep track of the directions of objects relative to the doll.

A second group of subjects responded to direction probes while observing the model scene. The doll was physically rotated and reclined in the model. Subjects could adopt the perspective of the doll in a viewed scene, but this would create a conflict between a spatial framework from the doll's perspective and the subject's own outside perspective. The other, more plausible, possibility is that subjects would use intrinsic computation to determine directions and report objects. This would eliminate the conflict between perspectives. In this case, subjects should be fastest to head/feet, followed by front/back, followed by left/right for *both* upright and reclining postures, in contrast to predictions of the spatial framework analysis.

The results of this experiment, shown in Table 13.3, are quite clear. When subjects responded from memory, the pattern of response times conformed to the spatial framework pattern. Critically, front/back was faster than head/feet for reclining dolls. This suggests that, as for narratives, when subjects observe a model of a person in a scene, they mentally

TABLE 13.3
Mean Response Times (in Seconds) for Memory and Perception
of a Model Scene With an Upright and Reclining Doll

| Posture | Direction | | | | | |
	Head	Feet	Front	Back	Left	Right
RESPOND FROM MEMORY						
Upright	1.80	1.84	2.02	2.27	2.67	2.73
Pairwise means		1.82		2.15		2.70
Reclining	2.58	2.60	2.28	2.31	3.13	3.24
Pairwise means		2.59		2.29		3.18
RESPOND FROM PERCEPTION						
Upright	1.01	1.00	1.07	1.08	1.25	1.28
Pairwise means		1.00		1.08		1.26
Reclining	1.12	1.12	1.20	1.21	1.36	1.39
Pairwise means		1.12		1.20		1.38

Note. From Bryant, Tversky, and Lanca (1997), Experiment 3. Copyright © 1997 by D. Bryant. Adapted by permission.

adopt that person's perspective and construct a mental spatial framework. They do so rather than employ their own external perspective, which would be relatively difficult and time consuming in this kind of memory task (see Franklin, Tversky, & Coon, 1992). When subjects responded while actually viewing the model scene, response times conformed to predictions of the intrinsic computation analysis. Critically, head/feet was faster than front/back for both upright and reclining dolls. These results indicate that subjects do not employ spatial frameworks during perception of a person in a scene.

When subjects learned scenes by observation, two consistent findings emerged. First, in both situations, when subjects responded from memory, the pattern of response times was identical to the spatial framework pattern found when subjects learn scenes from narratives. This implies that the spatial mental models constructed from observation are functionally equivalent to those constructed from descriptions. This is also evidence that the kinds of spatial regularities that underlie the spatial framework analysis are general, powerful determinants of memory for location. Second, the patterns of response times obtained when subjects observed the scene was considerably different from the patterns obtained when subjects responded from memory. This was true when subjects scanned scenes that surrounded themselves and when they viewed a model scene. Thus, spatial frameworks are not like internalized perceptions. The process of perception activates qualitatively different sets of spatial regularities that affect behavior.

2.3. Perceiving and Remembering Diagrams

Another way of dealing with space is by use of diagrams. This is an interesting case because a diagram is intermediate to language and physical environments. A diagram is representational, intended to convey spatial information about a place that is not physically present, just as language. A diagram, however, is also a physical thing having its own spatial properties, just as real environments. The study of diagrams has ecological justification in that maps, sketches, and pictures are commonly used to provide spatial information. They also allow us to determine what sort of mental framework is used to understand depictions of space. When people view a diagram depicting a spatial configuration, do they use the spatial concepts associated with language and memory or the spatial processes associated with perception?

In our recent studies (Bryant & Tversky, 1997), we considered perception of diagrams depicting a person surrounded by objects to all sides—a pictorial analog of the narratives and physical models studied before. An example is shown in Fig. 13.2. To avoid potential confounds of object

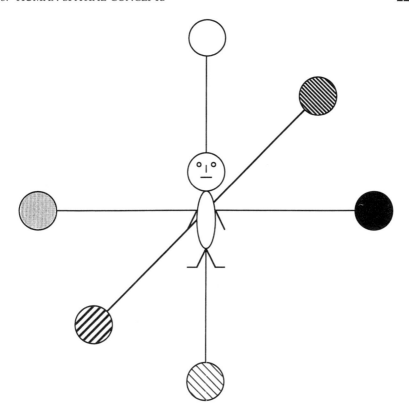

FIG. 13.2. From Bryant and Tversky (1997). Copyright © 1997 by D. Bryant. Adapted by permission. The diagonal line depicts relations in depth. In the experimental materials, each circle was presented in a unique color (blue, red, yellow, green, pink, or black). The configuration of colors around the person was random from trial to trial. The person was shown in one of four orientations, and within each orientation was rotated around its head/feet axis, so that orientation and facing were unpredictable from trial to trial.

identifiability, these diagrams used colored circles as targets and subjects named the color in response to direction probes. The question becomes, for this sort of diagram, do people construct spatial frameworks of the depicted spatial situation or do they use intrinsic computations to determine locations of objects? The former result would indicate that people impose highly conceptual representations of space on diagrams despite the physical spatial nature of diagrams. The latter result would indicate that perceptual processes predominate even when people view spatial figures that do not strictly conform to real 3-D environments.

Which result will occur is not immediately clear. Although perception of physical scenes invokes intrinsic computation, a diagram is unlike a real scene in several important respects. A diagram is two-dimensional and depth must be inferred by cues such as linear perspective. Diagrams can be held vertically, but are often viewed flat so that neither of its two dimensions is canonically aligned with gravity (i.e., a diagram has no fixed relation to gravity). The top of a diagram can also be rotated with respect to the viewer, so that the diagram's vertical does not correspond to the viewer's intrinsic vertical (i.e., a diagram has no fixed relation to the observer). These factors may make diagrams so abstract that a viewer would prefer to construct a spatial framework to understand the situation depicted.

In one experiment, subjects viewed 288 diagrams like that in Fig. 13.2. The orientation of the figure was varied within subject so that the person was upright, reclining to the left, reclining to the right, and upside-down on an equal number of trials. The order of diagrams was random. Subjects received a direction probe for each diagram and named the color at that direction for the person's perspective. The multiple orientations make it possible to distinguish between the spatial framework and intrinsic computation analyses. The former predicts faster response times to front/back than head/feet for a reclining person, but the latter predicts the reverse. The results shown in Table 13.4 are clear. At all orientations, subjects were faster to head/feet than front/back than left/right. Thus subjects employed intrinsic computation in perception of diagrams and we can rule out the hypothesis that subjects mentally adopt the perspective of a person depicted in a diagram. Even though the drawings were highly representational, they tapped the same perceptual spatial concepts that guide the locating of objects in observed 3-D physical scenes.

TABLE 13.4
Mean Response Times (in Seconds) for Perception
of Diagrams With a Person at Four Orientations

	Direction					
Orientation	Head	Feet	Front	Back	Left	Right
Upright	1.21	1.27	1.64	1.60	2.26	2.16
Pairwise means		1.24		1.62		2.21
Upside-down	1.26	1.29	1.87	1.82	3.16	3.41
Pairwise means		1.28		1.85		3.29
Reclining to left	1.27	1.31	1.86	1.80	2.32	2.32
Pairwise means		1.29		1.83		2.32
Reclining to right	1.33	1.28	1.86	1.78	2.49	2.43
Pairwise means		1.31		1.82		2.46

Note. From Bryant and Tversky (1997). Copyright © 1997 by D. Bryant. Adapted by permission.

Given that subjects apply the intrinsic computation analysis in perception of diagrams, a further question was whether spatial frameworks would be used in memory for diagrams. Viewers construct spatial frameworks for memory of observed 3-D scenes and narratives, suggesting that spatial frameworks are a general strategy for *remembering* spatial configurations. This makes sense because, regardless of how one has learned a scene, memory demands some form of mental reconstruction of the spatial array to respond to questions. Spatial frameworks are efficient memory representations for the sorts of scenes studied here. Only one perspective, that of the central person, is maintained in a spatial framework, which reduces cognitive load. Further, spatial frameworks can easily be updated as the person rotates or changes posture in the scene.

An experiment examined memory for scenes depicted in diagrams, again contrasting the spatial framework and intrinsic computation analyses for four orientations (upright, reclining to the left, reclining to the right, or upside-down). Although we had expected subjects to create spatial frameworks for memory of diagrams, we were proved wrong. In this experiment, subjects studied diagrams like that in Fig. 13.1, learning the positions of objects around a person in a particular scene. They then put the diagram aside and responded to direction probes on a computer. Table 13.5 shows that subjects' response times conform to predictions of the intrinsic computation analysis; that is, subjects were faster to head/feet than front/back at *all* orientations of the person.

This result is puzzling given previous findings that subjects create spatial frameworks for memory of observed environments and 3-D mod-

TABLE 13.5
Mean Response Times (in Seconds) for Memory
of Diagrams With a Person at Four Orientations

| Orientation | Direction | | | | | |
	Head	*Feet*	*Front*	*Back*	*Left*	*Right*
Upright	3.44	3.37	3.78	3.94	4.69	4.57
Pairwise means		3.40		3.86		4.63
Upside-down	3.80	3.80	4.06	4.35	6.25	6.05
Pairwise means		3.80		4.20		6.15
Reclining to left	3.94	4.02	4.12	4.36	4.79	4.80
Pairwise means		3.98		4.23		4.79
Reclining to right	3.99	4.09	4.09	4.11	4.82	4.67
Pairwise means		4.04		4.10		4.74

Note. From Bryant and Tversky (1997). Copyright © 1997 by D. Bryant. Adapted by permission.

els as well as narratives. Diagrams being like physical scenes and narratives in different respects were expected to also elicit spatial frameworks. Instead, subjects may have relied on images or perceptionlike representations of the diagrams and updated the images with reorientations of the character. Images are picturelike and simulate visual perceptual experience. Most important, they encode the whole diagram relative to the subject's outside viewpoint. When probed, subjects might mentally inspect their image as they would a visible diagram, analyzing the human person in terms of its major axes. Thus, as with visible diagrams, subjects first extract the head/feet dimension, followed by the front/back, and finally the left/right. Given the large body of evidence that operations performed in imagery are functionally equivalent to those in perception (see Finke & Shepard, 1986), such a strategy is consistent with the data. The results of memory for diagrams stand in sharp contrast to those of memory for narratives and physical scenes and indicate that subjects use a fundamentally different strategy.

3. CONCLUSION

When reading descriptions of scenes in which a person is surrounded by objects, people create spatial frameworks. Spatial frameworks are mental models that serve as a mental scaffolding on which information can be arranged and rearranged. The scaffolding is based on spatial concepts that define directions in terms of the body axes and enduring physical and perceptual regularities to which the mind has adapted. The processes by which a spatial framework serves as a mental scaffold may not be well understood, but clearly they are determined by the nature of the real world. Spatial frameworks are similarly used to remember observed physical scenes, either ones a person has actually been in or model scenes observed from outside.

People use a very different kind of framework when actually perceiving environments. Here, subjects employ intrinsic computation to locate objects around a person in an observed model scene or a diagram of a scene. This analysis is based on different spatial concepts than a spatial framework, ones related to the way people perceive objects. According to the intrinsic computation analysis, subjects extract the body axes of a person in a scene, then scan in the appropriate direction. Subjects are faster to identify objects at the head/feet than front/back in all orientations because people extract an object's vertical axis first when perceiving it.

Initially, it appeared that the use of spatial frameworks or intrinsic computation reflected a difference in representations for memory as opposed to perception. A conundrum raised by this research, however, is that subjects' performances in memory for diagrams contradict this formulation. Subjects were observed to use intrinsic computation even in memory for diagrams. This means that there is not simply a respond-from-memory versus respond-from-perception dichotomy, but also a dichotomy between learning from perception versus learning from description. One explanation is that subjects created imagelike representations of the diagrams in which the depicted scene is treated as an object in memory and coded with respect to the self. Further research will hopefully clarify this possibility.

In our research, we began with the general premise that the mind has internalized functional properties of human experience of space. Our approach has been to explore in detail the specific physical, perceptual, and behavioral regularities that define spatial concepts in one kind of situation. This has allowed us to make specific predictions and contrast two mental strategies for thinking about this spatial scene. From our experimental results, we have confirmed that human spatial concepts are defined not by just geometric properties, but functional properties of the self in space as well. Further, to understand how human cognition reflects the world we must consider the demands of the task, be it memory or perception, and the way one learns an environment, be it through language, models, or pictures. Thus, the internal worlds we create do not form maps of external space per se, but of perceptual and behavioral affordances within space.

ACKNOWLEDGMENTS

This research was sponsored by the Air Force Office of Scientific Research, Air Force Systems Command, USAF, under Grant or Cooperative Agreement number AFOSR 94-0220, and by a Research and Scholarship Development Fund grant from Northeastern University.

REFERENCES

Bryant, D. J., & Tversky, B. (1992). Assessing spatial frameworks with object and direction probes. *Bulletin of the Psychonomic Society, 30*, 29–32.

Bryant, D. J., & Tversky, B. (1997). *Computing spatial relations in diagrammed scenes.* Unpublished manuscript.

Bryant, D. J., Tversky, B., & Franklin, N. (1992). Internal and external spatial frameworks for representing described scenes. *Journal of Memory and Language, 31*, 74–98.

Bryant, D. J., Tversky, B., & Lanca, M. (1997). *Retrieving spatial relations from observation and memory.* Unpublished manuscript.

Clark, H. H. (1973). Space, time, semantics and the child. In T. E. Moore (Eds.), *Cognitive development and the acquisition of language* (pp. 26–63). New York: Academic Press.

Corballis, M. C., & Cullen, S. (1986). Decisions about the axes of disoriented shapes. *Memory & Cognition, 14*, 27–38.

Finke, R. A., & Shepard, R. N. (1986). Visual functions of mental imagery. In K. R. Boff, L. Kaufman, & J. P. Thomas (Eds.), *Handbook of perception and performance: Vol. 2. Cognitive processes and performance* (pp. 37:1–37:55). New York: Wiley.

Franklin, N., & Tversky, B. (1990). Searching imagined environments. *Journal of Experimental Psychology: General, 119*, 63–76.

Franklin, N., Tversky, B., & Coon, V. (1992). Switching points of view in spatial mental models acquired from text. *Memory & Cognition, 20*, 507–518.

Jolicoeur, P. (1985). The time to name disoriented natural objects. *Memory & Cognition, 13*, 289–303.

Jolicoeur, P., Ingleton, M., Bartram, L., & Booth, K. S. (1993). Top-bottom and front-behind decisions on rotated objects. *Canadian Journal of Experimental Psychology, 47*, 657–677.

Levelt, W. J. M. (1984). Some perceptual limitations on talking about space. In A. J. van Doorn, W. A. van de Grind, & J. J. Koenderink (Eds.), *Limits in perception* (pp. 328–358). Utrecht, The Netherlands: VNU Science Press.

Maki, R. H. (1986). Naming and locating the tops of rotated pictures. *Canadian Journal of Psychology, 40*, 368–387.

Miller, G. A., & Johnson-Laird, P. N. (1976). *Language and perception.* Cambridge, MA: Harvard University Press.

Rock, I., (1973). *Orientation and form.* New York: Academic Press.

Shepard, R. N., & Hurwitz, S. (1984). Upward direction, mental rotation, and discrimination of left and right turns in maps. *Cognition, 18*, 161–193.

14

How Addressees Affect Spatial
Perspective Choice in Dialogue

Michael F. Schober
New School for Social Research, New York

Spatial expressions are usually considered to reflect one of three possible perspectives: deictic, intrinsic, or extrinsic. I argue that because speakers' and addressees' viewpoints don't always coincide, we need more fine-grained distinctions. I partially describe two psychological experiments on how perspectives are chosen when viewpoints do not coincide. The results show that addressees' perspectives are indeed chosen often, and also that when people can, they often use spatial expressions that avoid the conflict in viewpoints. People also modify their spatial expressions as a result of their partner's conversational feedback of understanding. This suggests that the generation of spatial expressions includes a component that considers the addressee's viewpoint and what the addressee gives evidence of having understood.

1. INTRODUCTION

A fundamental question in the study of spatial expressions is how coordinate systems or perspectives come into play in their production and interpretation. An expression like "on the left" can describe a completely different spatial relation if it is interpreted from the speaker's point of view rather than from the point of view of an object—say, a chair—in the scene. This observation is not new: Cognitive psychologists (e.g., Levelt, 1982, 1984, 1989; Miller & Johnson-Laird, 1976), linguists (e.g., Fillmore, 1982; Lyons, 1977), and artificial intelligence (AI) researchers (e.g., Gapp, 1994;

Herskovits, 1986; Lang, Carstensen, & Simmons, 1991; Olivier, Maeda, & Tsujii, 1994; Retz-Schmidt, 1988) have all included perspective in their models of the representations and processes involved in spatial expressions.

Speakers are generally said to have two main perspective options in spatial descriptions: *deictic* and *intrinsic* perspectives. Some researchers include a third option: *extrinsic* perspective. Deictic terms are those whose interpretation depends on the speaker's viewpoint: left, right, front, back, and so on. Intrinsic descriptions use the point of view of some entity other than the speaker that has intrinsic parts—most often a front, back, left, and right—and a canonical orientation (see Lang et al., 1991, for a discussion of certain subtypes of the intrinsic perspective). Extrinsic descriptions describe an object's location by relying on other contextual factors, like the object's direction of motion, the accessibility of a reference object, or the earth's gravitation.

In this chapter I propose that in human conversations more perspectives may be at work than most researchers have distinguished. I describe the results of two psychological experiments on conversations about location. From analyses of the perspectives embodied in the locative expressions in these conversations, I argue that an analysis of which perspectives are easier or harder for people to take may predict which perspectives they prefer. I also argue that because spatial expressions are often embedded in conversational context, people can use the collaborative resources of conversation to interpret and disambiguate them. This has implications for how AI systems deal with descriptions they cannot immediately interpret.

2. HOW TO CLASSIFY SPATIAL PERSPECTIVES

One immediately obvious feature of human conversations about locations is that the speaker and the addressee don't always share the same vantage point. I use the term *addressee* rather than *listener* or *hearer* because listeners come in several different flavors. Listeners can be conversational participants who are currently being addressed (*addressees*), conversational participants currently not being addressed (*side-participants*), nonparticipants whose presence and attention are known to the participants (*bystanders*), or nonparticipants whose presence and attention are not known to the participants (*eavesdroppers*). Listeners in these different roles do not understand references in the same ways. For example, eavesdroppers do not understand as many references in conversations about novel topics as addressees, even when the eavesdroppers have heard every word the speaker and addressee have said to each other from the moment they first met (Schober & Clark, 1989).

This observation—that speakers and addressees sometimes don't share viewpoints—complicates the deictic/intrinsic/extrinsic distinction. How shall the addressee's point of view be classified? For most linguists and cognitive psychologists, deictic perspectives by definition refer only to the speaker's point of view. A description that takes the addressee's point of view is classified as an intrinsic use, in that the addressee, like other objects in a scene, has intrinsic parts and a canonical orientation. A different approach is taken by Herskovits (1986) and Retz-Schmidt (1988), who consider the addressee's point of view as a special case of the deictic perspective.

It is unclear to me that either approach is right. It seems wrong to classify the addressee's perspective as deictic: Speakers must be doing something other than being egocentric when they take the point of view of a partner whose vantage point differs from their own. It also seems wrong to classify the addressee's perspective as intrinsic: Speakers may conceive of their conversational partner's frame of reference differently from the frames of reference of nonhuman objects in the scene. In fact, research on conversational interaction (see, e.g., Clark & Marshall, 1981) raises the possibility that speakers model their addressees' knowledge and understanding, and that addressees therefore have privileged status in speakers' planning and production processes. (Interestingly, Schirra & Stopp, 1993, include a model of the listener's current and anticipated mental images as a necessary component of their SOCCER system, which generates radio-reporter-style descriptions of events in a videotaped soccer game.)

Another complication for researchers classifying spatial perspectives is what to do with ambiguous descriptions that can be classified as belonging to more than one perspective. For example, if a target item to be localized is "in front" for the speaker, but it also happens to be "in front" from the perspective of the chair in the scene, is this a deictic or intrinsic localization? As another example, how should the description "to the left" be classified when it is true from both the speaker's and the addressee's points of view, but those points of view are not identical? As observers we cannot decide. Speakers may have only one perspective—their own *or* their addressee's—in mind. Or they may have a combined or shared perspective in mind—they may conceive of the vantage point as "ours" for the moment. This raises the possibility that ambiguous perspectives may actually be perspectives in their own right.

A final complication: Some descriptions don't belong in either the deictic or intrinsic categories, because they are true no matter what the speaker's *or* the addressee's points of view are. And they don't quite seem to belong in the extrinsic category, because they do not rely on the kinds of contextual factors Retz-Schmidt (1988) mentions. These include descriptions like "near the lamp," "between the car and the house," and "in the

middle of the room." Such descriptions have been described in the psychological literature as "local references without a coordinate system"
(Levelt, 1989), descriptions "ohne erschliessbare Origobesetzung," that is,
without a recoverable frame of reference (Herrmann, Dietrich, Egel, &
Hornung, 1988), topological localizations (Egel & Carroll, 1988), and descriptions in a "landmark" frame of reference (Craton, Elicker, Plumert,
& Pick, 1990; Pick, Yonas, & Rieser, 1979). But the psychological processes
involved in their use have received little attention.

I propose, then, that we should use at least the following perspective
categories to describe localizations in human conversations: (a) speaker
centered (traditionally called deictic or egocentric), (b) addressee centered
(traditionally classified as intrinsic), (c) "both" centered (ambiguous with
respect to speaker's or addressee's perspective), (d) object centered (traditionally classified as intrinsic), (e) extrinsic (environment centered), and
(f) neutral (with respect to frame of reference). Other kinds of possible
ambiguities may constitute perspectives in their own right as well:
speaker-object, addressee-object, and speaker-addressee-object. Of course,
not all these perspectives are available to speakers in every situation;
characteristics of scenes come into play.

3. PSYCHOLOGICAL STUDIES OF SPATIAL
PERSPECTIVES IN CONVERSATION

Very little psychological research on spatial perspectives has examined
situations where the speaker's and addressee's viewpoints differ. Most
work has assumed that speaker and addressee share the same vantage
point. One exception is the work of Herrmann and his colleagues; they
have carried out a number of studies in which speakers have imaginary
addressees at different vantage points (e.g., Bürkle, Nirmaier, &
Herrmann, 1986; Herrmann, 1989; Herrmann, Bürkle, & Nirmaier, 1987).
But in none of those studies have speakers had an actual live, interacting
addressee who cared to know what the speaker intended to express.

In contrast, my studies explore which perspectives people take when
their (live, interacting) conversational partners' vantage points differ. Here
I describe some relevant results from two of these studies.

3.1. Study 1: Spatial Perspectives in Conversation
Versus Monologue

This study (Schober, 1993) compared how people choose perspectives
when they have an actual partner and when they have the usually studied
imaginary partner. Subjects were required to describe locations on a series
of simple displays for a partner whose viewpoint switched from display

to display. Because of the nature of the displays, speakers had to take one of three perspectives: speaker-centered, addressee-centered, or both-centered perspectives.

In the study, subjects I call *directors* viewed a series of 32 simple spatial displays. Directors were separated by a visual barrier from their conversational partners, the *matchers*, so that they would have to speak in order to communicate locations. Matchers viewed a series of displays that corresponded to the directors'. The displays contained two small circles within a large circle. The two small circles were positioned either vertically, horizontally, diagonally from bottom left to top right, or diagonally from top left to bottom right. On each display, one small circle was designated the *target* circle. The directors' task was to convey to the matchers which circle was the target, so that the matchers could mark the target on their own display.

At the bottom of the page (the side closest to the viewer) was an arrow and the word *you*, showing the viewer's vantage point. Another arrow along with the word *partner* showed the other person's viewpoint. The two partners' vantage points were either (a) the same, (b) off by 90°, (c) off by 180°, or (d) off by 270°. Figure 14.1 shows a set of sample displays.

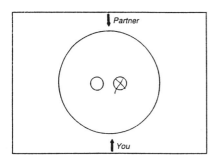
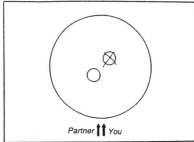

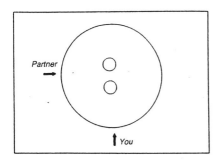
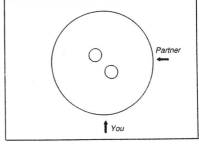

FIG. 14.1. Study 1 sample displays. From Schober (1993). Copyright © 1993 by Elsevier Science Publishers. Reprinted by permission.

Some directors described locations into a tape recorder for an imaginary partner, just as in the Herrmann studies. Other directors had actual matchers present (other naive subjects) whose task was to mark the target circles on their own series of displays. Of course, on the matchers' displays "You" referred to the matcher's own viewpoint, and "Partner" referred to the director's viewpoint. All subjects were encouraged to say anything they needed to in order to accomplish the task. This generated a set of dialogues and monologues, that were then carefully transcribed.

Locative expressions in the transcripts were identified and coded for perspective (speaker-centered, addressee-centered, or both-centered). These locative expressions were mostly phrases containing projective or topological prepositions and adjectives, but there was a wide variety of them. Speakers most often used *left, right, front, behind,* but they also used *above, below, upper, lower,* and *back.* Also popular were descriptions like *closest, farthest,* and *nearest.* Not surprisingly for those who observe natural speech, some descriptions were nongrammatical (though sensible), like "to the rightmost of you." Also notable is that speakers often used more than one locative expression to describe a single location, and these multiple locatives did not always reflect only one perspective.

3.1.1. Results. On average, directors in this study most often took the matcher's perspective. But there were striking differences between the speakers with actual partners and the speakers with imaginary partners. The speakers with imaginary partners almost never used their own perspective; they almost exclusively took their partners' perspectives. In contrast, speakers with actual live partners were significantly more likely to be egocentric. In fact, some were egocentric 100% of the time.

Why should this be? Intuitively, one might expect that speakers with real partners should be *less* egocentric, because the presence of a real partner might make a person more aware of the difference in perspective. But I believe these results make sense if one considers how spatial perspectives operate in a conversational context, and how differently they operate in a monologue. In conversation, one of people's primary goals is to be understood. Speakers monitor carefully for evidence that they have been understood (Brennan, 1990/1991, 1996; Clark & Wilkes-Gibbs, 1986). When speakers have been understood, they are licensed to continue with their current descriptive strategy; when they get evidence that they have not been understood, they ought to switch to another descriptive strategy in hopes that it may work. In Study 1, speakers who used egocentric descriptions in dialogue could get feedback that being egocentric was a successful strategy (i.e., their addressees were able to understand them without difficulty), and so they were licensed to continue being egocentric. In monologues, on the other hand, where speakers did not get immediate evidence of whether they had been understood, speakers could not know whether an egocentric strategy was going to be successful. The reasonable thing for speakers to do was to take their partner's perspective exclusively, because that was almost sure to be understood.

This interpretation of the results is supported by an analysis of the small subset of conversational interchanges where the director's first description of a target location on one display was not understood right away. In these interchanges, the misunderstanding resolved in one of two ways: (a) The matchers ended up understanding and verbally accepting the director's initial perspective, or (b) the matchers rejected the director's initial perspective, either proposing a different perspective or requiring the director to propose a different perspective, and gave evidence of understanding this different perspective. In the first case, directors continued using the original perspective—the one the addressee gave evidence of having understood—in their subsequent descriptions. But in the second case, directors changed strategies in their subsequent descriptions, no longer proposing the problematic perspective.

This study's results suggest, then, that interactive feedback of understanding from a conversational partner shapes the way speakers generate spatial expressions.

3.2. Study 2: Conversational Perspectives in Descriptions of a Complex Display

This study (Schober, 1995) examined how conversing pairs of subjects negotiate perspective when describing locations on a much more complicated display that allowed subjects to take any of five perspectives: speaker-centered, addressee-centered, both-centered, object-centered, or neutral perspectives. Unlike in the earlier study, these subjects described locations on only this one complex display, and their partner's viewpoint was fixed throughout the study. All pairs of subjects were again separated by a visual barrier so that they could only communicate verbally; again, in each pair one subject was designated the director and one the matcher. The director and matcher each viewed an identical display on either side of the barrier. Some directors had matchers who shared their vantage point (who faced the display from the same viewpoint), others had matchers whose vantage point differed from theirs by 90°, and others had partners whose vantage point differed from theirs by 180°.

The display was a large white circular sheet of cardboard (50 cm in diameter) with identical red paper circles, blue paper triangles, and purple paper squares on it (see Fig. 14.2). The shapes were in a configuration that had no obvious pattern or design. The display had eight of each kind

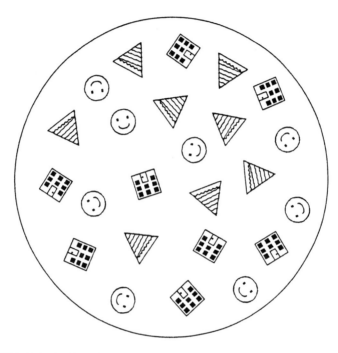

FIG. 14.2. Study 2 display. From Schober (1995). Copyright © 1995 by
Ablex Publishing Corporation. Reprinted by permission.

of shape, so that subjects would be forced to describe locations rather
than simply identify shapes. Each shape had a design drawn on it. These
were the designs: All the circles had the identical rudimentary face drawn
on them, all the triangles had the identical diagonal design (so that one
side was distinctive), and all the squares had identical windows and doors
drawn on them, in a house pattern. Unexpectedly, the display ended up
looking like a large pizza with unusual toppings.

The task was this: An experimenter shuffled a set of transparent plastic
chips with the number 1 to 18 on them, along with six blank chips, and
laid them out on the 24 shapes on the director's display. The chips were
transparent so that subjects could still see the designs on the shapes when
the chips were on the board. The experimenter explicitly told each subject
what the other's vantage point was, and showed it by pointing. Directors
were to get their matchers to lay their own set of plastic chips onto the
identical shapes on their own display. They were to start with chip number
1 and end with chip number 18. As in the last study, directors and
matchers were encouraged to say anything they wanted to each other to
accomplish the task. The 27 pairs of subjects performed this task three
times. After two rounds where one subject was the director and the other
the matcher, they switched roles for a third round (the director became

the matcher, and vice versa). This generated an extensive set of complex dialogues, which were then transcribed.

3.2.1. Coding.
Transcripts of the conversations were again segmented into locative expressions, but the descriptions were much more complicated this time. The 5,873 descriptions that conveyed location in the corpus were identified. There were three types: locative phrases, directional phrases, and time locatives. Locative phrases were by far the most common. See Table 14.1 for details of the segmentation.

Each locative was coded as belonging to one of these five perspectives: director centered, matcher centered, both centered, object centered, or neutral. Locatives were coded as director and matcher centered rather than as speaker and addressee centered because the roles "speaker" and "addressee" constantly shift throughout a conversation. If we were to code the locatives according to these roles, we would miss the fact that when one partner says "on my left" and the other replies "yes, on your left" the same *person's* perspective has been used. To avoid confusion, all analyses paid strict attention to which partner had uttered which locatives; thus a director-centered locative uttered by the director is speaker cen-

TABLE 14.1
Segmentation of Locatives

---

(1) *Locative phrases* (80% of coded descriptions)

These appear in the form *Object A is (locative relation) object B* (or a close variant), where the locative relation was a string like "at the right of," "toward the left of," "southwest of," or "across from." Object B was most often a shape on the game board (terms like *square, house, circle, face,* or *triangle*), but it also sometimes represented a region ("the northeast quadrant") or the board itself ("the pizza"). Thus, a description like "on the left half of the pizza above the house" was counted as two locative expressions: "on the left half of the pizza" and "above the house." Only expressions that described relative location were coded, so expressions like "on a triangle" in the statement "put chip number three on a triangle" were not coded. Locatives that were only partly complete were not coded.

---

(2) *Directional phrases* (18% of coded descriptions)

These describe directions of motion in the form *From Object A go (directional phrase), passing object B, object C etc., on the way to final target T* (or a close variant). All the arguments are optional. The directional phrase is a string like "up," "west," "to your left," or "across the board." This directional phrase embodies a perspective in exactly the same way as the more typical locative expressions do.

---

(3) *Time locatives* (2% of coded descriptions)

These describe spatial adjacency by using temporal contiguity (e.g., "and then there's a house"). These were counted as locative expressions only in the absence of locative expressions that explicitly described spatial relations, as in "First you have a purple house and then a red smiley face," and they were relatively rare in the transcripts.

---

tered, and a director-centered locative uttered by the matcher is addressee centered.

Some locative phrases that seemed at first to be object centered (they explicitly described parts of shapes on the display) could not be coded as object centered. For example, a director's description like "the triangle whose point is pointing toward your left" could only be coded as matcher centered, because "left" only makes sense from the matcher's vantage point. Time locatives could only be coded as embodying a neutral frame of reference; there were few enough of these that this couldn't materially affect the analyses.

A few descriptions involving true object-centered perspectives could not be neatly assigned to one of the five perspective categories, because the object's own left, for example, corresponded to the speaker's or addressee's left. These were tagged as ambiguous. Nine locative expressions (out of the 5,873 coded) could not be assigned to any of the five categories, because they weren't true from any perspective, or were only true from a perspective that didn't correspond to either the director's or matcher's vantage point. These were left out of subsequent analyses.

The descriptions directors gave for each location were quite complex, consisting of multiple locatives. Directors averaged 3.79 locatives per figure in Round 1, 2.92 in Round 2, and 2.90 in Round 3. Matchers were often active in providing locative expressions as well, averaging 1.13 per figure in Round 1, .60 in Round 2, and .76 in Round 3. Even the simplest descriptions consisted of many conversational turns. Also, description strategies varied tremendously. Some directors used a clock metaphor, describing shapes as being located "at three o'clock" and the like. Others divided the pizza into quadrants, or concentric circles, or groupings of shapes ("the three red smiley faces in a row"). Some used a particular shape (often the center square) as a reference point they could go back to again and again. Many described the locations of nearby shapes in order to locate the target shapes; in fact, on 44% of the descriptions directors took their matchers on tours, stepping along shape by shape to their final destination.

3.2.2. Results. Some striking findings emerged from detailed analysis of the perspectives used in this corpus of conversation. These are four:

1. Unlike in the last study, directors very rarely used egocentric descriptions. They also used virtually no unambiguously object-centered perspectives. Instead, they used mostly addressee-centered, both-centered, and neutral perspectives. When matchers spoke, they used the corresponding perspectives. That is, where directors used addressee-centered perspectives (took their partner's perspective), matchers used

speaker-centered perspectives (took their own perspectives) in almost exactly the same proportions.

2. Partners who shared vantage points used mostly both-centered and neutral perspectives; partners with offset vantage points (90° and 180°) used mostly addressee-centered and neutral perspectives. Partners with offset vantage points used a higher proportion of neutral perspectives than shared-vantage-point partners. But there was virtually no difference in how 90°- and 180°-offset partners used perspectives.

3. Over time, partners used more and more neutral descriptions, and fewer and fewer both-centered descriptions. This was true within a round of 18 descriptions, and it was also true across rounds: partners used more neutral descriptions in Round 2 than in Round 1. This can't be attributed merely to display characteristics, because the display was the same in Round 2 as in Round 1.

4. When matchers became directors in Round 3, they used virtually the same proportions of the different perspectives (speaker centered, addressee centered, both centered, object centered, and neutral) as their partner had used in Round 2. In other words, when roles switched, subjects in this experiment did not stick with taking the perspective of the person whose perspective had been used most in the previous round. Rather, they opted for being egalitarian, taking their partner's perspective just as much as their partner had been taking theirs.

What explains these results? I propose that speakers may try to choose perspectives that minimize their own or their partner's effort (see Schober, 1995). In producing spatial expressions, speakers find it easier to speak egocentrically, and harder to take their (offset) addressee's perspective (Bürkle et al., 1986; Schober & Bloom, 1995). Correspondingly, it should be easiest for (offset) addressees to comprehend addressee-centered descriptions and harder to comprehend speaker-centered descriptions. In general, taking a perspective that requires speakers to shift frames of reference away from their own egocentric frame should be harder than taking any perspective that allows them not to.

On this account, the person-centered perspectives (speaker and addressee centered) can cause a collaborative problem for conversational partners, because they require people to choose one person's perspective. Both-centered perspectives don't cause a conflict. Object-centered perspectives may vary in difficulty, depending on the orientation of the object in question relative to the speaker or addressee. And neutral perspectives should be easy for speakers to produce and addressees to comprehend, because neither party has to shift coordinate systems away from their own.

If this analysis is correct, then what people did in this study reflects a strategy of minimizing effort. The strong preference for neutral descriptions when vantage points were offset and the increasing tendency to use neutral perspectives over time both suggest that partners were avoiding the collaborative difficulties that person-centered descriptions create. When people did use person-centered descriptions, they overwhelmingly chose the addressee's perspective, suggesting that it was the addressee's effort they were trying to minimize. This was true even when roles in the experimental task shifted; new directors used the addressee's perspective just as much as their partners had when they were directors.

4. IMPLICATIONS

I believe the results of these studies have certain implications for theories of spatial expression. First, the fact that speakers and addressees' vantage points can differ is a real issue: Speakers *do* take the addressee's perspective, rather than their own egocentric perspective, quite often when given the opportunity. To do this speakers must mentally represent their conversational partner's point of view, and this mental representation plays a role in the planning of spatial expressions. The addressee's point of view may be part of the speaker's mental representation even when it doesn't differ from the speaker's: Perhaps the very act of speaking necessitates a model of the partner that includes perspective similarities and differences.

Second, people use neutral descriptions that don't require them to take any one person's perspective a substantial amount of the time in interactive conversation—more than half the time, on average, in the conversations in Study 2. The use of these descriptions also increased over time. Speakers don't avoid person-centered descriptions entirely, but they are more likely to use neutral descriptions when points of view don't coincide. The processes involved in taking the addressee's perspective and in avoiding person-centered perspectives have not been investigated much, and they deserve more attention. Some preliminary results (Schober & Bloom, 1995) suggest that nonegocentric descriptions are not all equally difficult to produce.

Third, the complexity of the location descriptions in these studies is notable. People left to their own devices will use astonishingly complex, multifaceted description strategies in dealing with location, and their strategies go well beyond what most semantic and linguistic theories of spatial localization describe.

Fourth, conversational feedback of a partner's understanding may be more central to the processing of spatial expressions than we usually consider. In these studies, speakers modified their perspective choices and thus spatial expressions when they got evidence they weren't being understood. This suggests that the generation and planning of spatial expressions includes a model of a partner's understanding, a model that can be updated based on new input from the partner.

5. CONCLUSIONS

In many cases in the real world, and in a number of current AI systems that deal with spatial expression, speaker and addressee share the same viewpoint, and so the addressee's distinct viewpoint doesn't seem to be at issue. This may in fact be what motivates some researchers not to distinguish speaker- and addressee-centered descriptions. But the addressee's viewpoint is important in both theoretical and applied ways. For a general theory of spatial expression, it may be that the addressee's viewpoint is *always* represented, even when it coincides with the speaker's. Spatial expressions are used for the purpose of being understood by agents with communicative purposes, and the negotiation of that understanding plays an important role in both the production and comprehension of spatial expressions.

As for AI system applications, I raise three points relevant to both the generation and interpretation of spatial expressions:

1. Spatial communication must sometimes occur in situations where speaker and addressee do not share the same viewpoint—think of multiuser virtual spaces, or direction-giving applications. Perspective choice is a more complicated issue in these cases than in many current systems.

2. In the studies reported here, human conversational partners were highly egalitarian in their perspective choices: They reciprocated taking the other person's perspective just as much as the other had taken theirs. It is an open question what the optimal relationship for systems and users is.

3. The task of a spatial representation has been characterized as generating the ideal spatial expression or figuring out the most plausible interpretations of a spatial expression. When perspective ambiguities arise, some systems (e.g., Olivier et al., 1994) generate parallel interpretations in both perspectives. This may or may not be what people do in such situations (see, e.g., Carlson-Radvansky & Irwin, 1994). Either way, given the data I have described on how humans negotiate understanding of spatial expressions, I propose that rather than (or in addition to) trying to generate the perfect expression or create perfect interpretations, we might build in an architecture for accepting understanding and repairing misunderstandings. Human partners use conversational resources—ask-

ing questions, testing hypotheses, giving feedback of understanding—to disambiguate problematic spatial perspectives. Perhaps systems applications should be given similar resources, such that the burden of disambiguation rests more in the dialogue, as Brennan and Schober (1993) and Brennan and Hulteen (1995) have proposed in other domains.

REFERENCES

Brennan, S. E. (1990/1991). Seeking and providing evidence for mutual understanding (Doctoral dissertation, Stanford University, 1990). *Dissertation Abstracts International, 51*(11), 5611B.

Brennan, S. E. (1996). *Conversation on-line: The time course of coordinating beliefs.* Manuscript submitted for publication.

Brennan, S. E., & Hulteen, E. A. (1995). Interaction and feedback in a spoken language system: A theoretical framework. *Knowledge-Based Systems, 8,* 143–151.

Brennan, S. E., & Schober, M. F. (1993). *Speech disfluencies in spoken language systems: A dialog-centered approach.* Grant proposal to National Science Foundation, Interactive Systems Division.

Bürkle, B., Nirmaier, H., & Herrmann, T. (1986). *"Von dir aus . . .": Zur hörerbezogenen lokalen Referenz* ["From your point of view . . .": On listener-oriented local reference] (Bericht Nr. 10). Mannheim: University of Mannheim, Forschergruppe "Sprechen und Sprachverstehen im sozialen Kontext."

Carlson-Radvansky, L. A., & Irwin, D. E. (1994). Reference frame activation during spatial term assignment. *Journal of Memory and Language, 33,* 646–671.

Clark, H. H., & Marshall, C. R. (1981). Definite reference and mutual knowledge. In A. K. Joshi, B. Webber, & I. A. Sag (Eds.), *Elements of discourse understanding* (pp. 10–63). Cambridge, England: Cambridge University Press.

Clark, H. H., & Wilkes-Gibbs, D. (1986). Referring as a collaborative process. *Cognition, 22,* 1–39.

Craton, L. G., Elicker, J., Plumert, J. M., & Pick, H. L., Jr. (1990). Children's use of frames of reference in communication of spatial location. *Child Development, 61,* 1528–1543.

Egel, H., & Carroll, M. (1988). *Überlegungen zur Entwicklung eines integrierten linguistischen und sprachpsychologischen Klassifikationssystem für sprachliche Lokalisationen* [Considerations for the development of an integrated linguistic and psycholinguistic classification system for verbal localization] (Bericht No. 18). Mannheim: University of Mannheim, Forschergruppe "Sprechen und Sprachverstehen im sozialen Kontext."

Fillmore, C. J. (1982). Towards a descriptive framework for spatial deixis. In R. Jarvella & W. Klein (Eds.), *Speech, place, and action* (pp. 31–59). Chichester, England: Wiley.

Gapp, K.-P. (1994). Basic meanings of spatial relations: Computation and evaluation in 3D space. In *Proceedings of AAAI-94* (pp. 1393–1398). Menlo Park, CA: AAAI Press.

Herrmann, T. (1989). *Partnerbezogene Objektlokalisation—Ein neues sprachpsychologisches Forschungsthema* [Partner-oriented localization of objects—a new psycholinguistic research topic]. Mannheim: University of Mannheim, Forschergruppe "Sprechen und Sprachverstehen im sozialen Kontext."

Herrmann, T., Bürkle, B., & Nirmaier, H. (1987). Zur hörerbezogenen Raumreferenz: Hörerposition und Lokalisationsaufwand [On listener-oriented spatial reference: Listener position and localization effort]. *Sprache & Kognition, 3,* 126–137.

Herrmann, T., Dietrich, S., Egel, H., & Hornung, A. (1988). *Lokalisationssequenzen, Sprecherziele und Partnermerkmale: Ein Erkundungsexperiment* [Localization sequences, speaker goals, and partner features: An exploratory study]. Mannheim: University of Mannheim, Forschergruppe "Sprechen und Sprachverstehen im sozialen Kontext."

Herskovits, A. (1986). *Language and spatial cognition: An interdisciplinary study of the prepositions in English.* Cambridge, England: Cambridge University Press.

Lang, E., Carstensen, K.-U., & Simmons, G. (1991). *Modelling spatial knowledge on a linguistic basis* (Lecture Notes in Computer Science, Vol. 481). Berlin: Springer-Verlag.

Levelt, W. J. M. (1982). Cognitive styles in the use of spatial direction terms. In R. J. Jarvella & W. Klein (Eds.), *Speech, place, and action* (pp. 251–268). Chichester, England: Wiley.

Levelt, W. J. M. (1984). Some perceptual limitations on talking about space. In A. van Doorn, W. van de Grind, & J. Koenderink (Eds.), *Limits of perception: Essays in honour of Maarten A. Bouman* (pp. 323–358). Utrecht, The Netherlands: VNU Science Press.

Levelt, W. J. M. (1989). *Speaking: From intention to articulation.* Cambridge, MA: MIT Press.

Lyons, J. (1977). *Semantics* (Vol. 2). Cambridge, England: Cambridge University Press.

Miller, G. A., & Johnson-Laird, P. N. (1976). *Language and perception.* Cambridge, MA: Harvard University Press.

Olivier, P., Maeda, T., & Tsujii, J. (1994). Automatic depiction of spatial descriptions. In *Proceedings of AAAI-94* (pp. 1405–1410). Menlo Park, CA: AAAI Press.

Pick, H. L., Yonas, A., & Rieser, J. J. (1979). Spatial reference systems in perceptual development. In M. H. Bornstein & W. Kessen (Eds.), *Psychological development from infancy* (pp. 115–145). Hillsdale, NJ: Lawrence Erlbaum Associates.

Retz-Schmidt, G. (1988). Various views on spatial prepositions. *AI Magazine, 9*(2), 95–105.

Schirra, J. R. J., & Stopp, E. (1993). ANTLIMA—A listener model with mental images. In *Proceedings of IJCAI-93* (pp. 175–180). San Mateo, CA: Morgan Kaufman.

Schober, M. F. (1993). Spatial perspective-taking in conversation. *Cognition, 47,* 1–24.

Schober, M. F. (1995). Speakers, addressees, and frames of reference: Whose effort is minimized in conversations about locations? *Discourse Processes, 20,* 219–247.

Schober, M. F., & Bloom, J. E. (1995, November). *The relative ease of producing egocentric, addressee-centered, and object-centered spatial descriptions.* Poster session presented at the 36th annual meeting of the Psychonomic Society, Los Angeles.

Schober, M. F., & Clark, H. H. (1989). Understanding by addressees and overhearers. *Cognitive Psychology, 21,* 211–232.

15

Spatial Prepositions, Functional Relations, and Lexical Specification

Kenny R. Coventry
University of Plymouth

In this chapter a minimally specified approach to the lexical entries for (topological[1]) spatial prepositions is presented based on functional relations. To begin with, it is argued that fully specified accounts to the semantics of spatial terms manifest no advantages over minimally specified accounts, and furthermore that previous approaches in this domain, whether fully or minimally specified, are fundamentally flawed in that they conflate lexical concepts and categories. Functional relations are introduced as an alternative way of understanding spatial relations (Coventry, 1992, 1995; Garrod & Sanford, 1989; Talmy, 1988). It is argued that what is important about objects is how they interact with each other, that is, the *functional relations* between objects. A battery of empirical studies are reviewed that demonstrate that spatial language use is underdetermined by geometric relations in the input, and better predicted by functional relations reliant on object-specific knowledge. Given these data, a framework to the encoding and decoding problems is outlined encompassing mental models as an interface between language and the spatial world.

---

[1]This chapter does not deal explicitly with projective prepositions, which have been given a more adequate treatment in the present author's view than so-called topological prepositions. For a detailed consideration of projectives see Coventry and Garrod (in preparation), or Retz-Schmidt (1988) or Herskovits (1986) for a review.

1. INTRODUCTION

Approaches to the meaning of spatial prepositions have largely assumed that spatial prepositions define a region in which the figure (object to be located) is located with respects to the ground (reference object). For instance, in the sentence, "The pear is in the bowl," the figure (*the pear*) is located in the region described by the prepositional phrase *in the bowl*, with the spatial relation expressed by *in* corresponding to something like "contained interior to the reference object." Thus the content the spatial preposition contributes to the sentence has to do with the type of spatial relation defining the region, and for those interested in the lexical seman-tics of spatial terms, the task has been one of identifying the specific regular content(s) associated with such terms (a necessary goal if one is to explain how one produces and understands novel utterances).

The problem is that there are a large number of spatial relations that are appropriate for each term. For example, the preposition *in* is appro-priate to describe the relationship between the various figures and grounds depicted in Fig. 15.1; therefore, specifying a single content for *in* that is appropriate for all these situations is problematic. To complicate matters further, there is also not a one-to-one mapping between spatial relations and prepositional usage, as the cases in Fig. 15.2 demonstrate. For example, Fig. 15.2(a) is the same spatial relation as Fig. 15.1(a), but

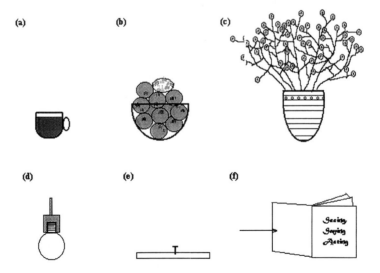

FIG. 15.1. Different spatial relations for which *in* is appropriate: (a) The coffee is *in* the cup, (b) The lemon is *in* the bowl, (c) The flowers are *in* the vase, (d) The lightbulb is *in* the socket, (e) The nail is *in* the board, (f) The page is *in* the book.

(a)   (b)   (c)

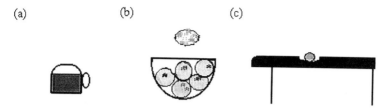

FIG. 15.2. Different spatial relations for which *in* is not appropriate: (a) *The coffee is *in* the cup, (b) *The lemon is *in* the bowl, (c) *The ball is *in* the table.

this time *in* is not appropriate. Geometrically, Fig. 15.2(b) is the same as Fig. 15.1(b), but again *in* is not appropriate. Similarly, Fig. 15.2(c) is the same again as Fig. 15.1(c), but again *in* is not appropriate. In essence there are two types of case accountability problems that one must be aware of. The first is the case where the lexical entry fits situations where the preposition is not appropriate (there are *decoding* errors), and the second is where the lexical concept used simply does not fit the data (these are *encoding* errors).

This has forced researchers to make the choice between minimal or full specification of lexical entries. Either one can put a minimal content in the lexicon for the preposition (as was done by Cooper, 1968; Leech, 1969; Lindkvist, 1950; Lindner, 1981; Miller & Johnson-Laird, 1976; Sandhagen, 1956) and give an account as to how the content can be bent and stretched (using pragmatics), or alternatively one can fully specify the lexical entry (as was done by Bennett, 1990; Brugman, 1981, 1988; Brugman & Lakoff, 1988; Casad, 1982; Hawkins, 1984; Herskovits, 1985, 1986; Janda, 1984; Lakoff, 1987; Rudzka-Ostyn, 1983), and give an account as to how a particular sense is selected in context. In recent years fully specified accounts have been by far the most popular approaches adopted, often within the framework of cognitive linguistics. This mirrors the move in the concept literature from so-called classical[2] theories of concepts to probabilistic theories, most notably prototype theories.

In this chapter it is argued that there are no grounds a priori for preferring fully specified accounts to minimally specified accounts for the lexical semantics of spatial prepositions. More important, it is argued that previous fully and minimally specified approaches to spatial prepositions are fundamentally flawed for the same reason, namely they confuse lexical

---

[2]The term *classical* is used here as used by Lakoff (1987). In reality such theories may be something of a straw man—traditional lexicographers were well aware of the problems of polysemy and homonymy, as well as phenomena such as that of metonymy. We therefore treat the classical theory as an abstract and somewhat extreme view, which is useful as a reference point for a consideration of theories of word meaning.

concepts with categories. In particular, the lexical concepts used have to do with spatial relations in the world. As these are difficult to characterize in terms of categories (Crangle & Suppes, 1989; Suppes, 1991), they are unlikely to be mentally representable. An alternative theory of the concepts associated with spatial prepositional usage is presented that focuses on functional relations, which have to do in a real sense with what cognitive linguists have termed *experientialism*. In essence, it is argued that categories indeed have to do with spatial relations in the world, but that lexical concepts have to do with experience of that world. Furthermore, empirical evidence is provided that demonstrates that spatial prepositional usage is underdetermined by spatial relations alone, and better predicted by functional relations, thus supporting the view provided. Finally the implications for artificial intelligence (AI) machine translation systems and interfaces are discussed.

2. FULL VERSUS MINIMAL SPECIFICATION FOR LEXICAL ENTRIES

One can address the issue a priori as to whether there are any grounds for favoring full or minimal specification for lexical entries. To begin with, it is clear that word meaning, at least in extension, is extremely variant. For example, if we take a spatial preposition like *in*, there appear to be a large number of spatial relations for which this term is appropriate (a selection of which are displayed in Figs. 15.1 and 15.2). This leads to the issue of whether one should attempt to find something common to all occurrences of a lexeme, and make the assumption of *Gesamtbedeutung* ("general meaning"; cf. Jakobson, 1932, 1936), or whether one should adopt fully specified lexical entries recognizing many senses.

Lakoff (1987) is one of the strongest critics of what he termed the "classical" approach to lexical concepts and categories. His arguments against minimally specified lexical entries revolve around two main criticisms, represented in the following quote:

> The classical theory of categories does not do very well on the treatment of polysemy. In order to have a single lexical item, the classical theory must treat all of the related senses as having some abstract meaning in common—usually so abstract that it cannot distinguish among the cases and so devoid of real meaning that it is not recognisable as what people think of as the meaning of a word. And where there are a large number of related senses that don't all share a property, then the classical theory is forced to treat such cases as homonymy, the same way it treats the case of the two words *bank*. Moreover, the classical theory has no adequate means of char-

acterising the situation where one or more senses are "central" or "most representative." (p. 416)

Breaking this down, Lakoff was making the claim that fully specified lexical entries should be adopted so that prototype effects can be adequately encapsulated, and so that the relationship between senses can be preserved thus avoiding homonymy. We can take each of these claims in turn.

The existence of prototype effects is not in question. Originally, Rosch and colleagues (Rosch, 1973, 1975, 1977; Rosch & Mervis, 1975) found, across a variety of experimental paradigms, that some examples of a category are more typical or central than others. However, although the effects are not in question, their origins and representation are. Despite early claims made by Rosch (1973, 1975) that prototype effects provide a characterization of the internal structure of a category in the mind, and that prototypes constitute mental representations, there is much evidence to suggest that prototypes are merely effects, not representations.

There are many cases where typicality effects occur in categorization judgments where it can be demonstrated that subjects cannot have this information stored a priori as the category is an ad hoc one. Barsalou (1983) found that prototype effects exist for categories such as *things to take on a camping trip* and *things to do on a wet weekend*. If effects exist for categories that clearly are not stored in advance, then it is likely, *ceteris paribis*, that they are not directly represented for stored categories.

Not only do prototype effects exist for ad hoc categories, but they also exist for well-defined categories. Armstrong, Gleitman, and Gleitman (1983) found that subjects readily order instances according to their perceived typicality despite knowing full well that the category has sharp boundaries, that there are clear conditions of membership, and that membership itself is either all or none. This indicates the fallacy in assuming that prototype effects are a direct reflection of graded membership.

Further evidence against the direct representation of prototypes comes from the manipulation of context in various ways. For example, Roth and Shoben (1983) based a study on a finding of Garrod and Sanford (1977), from which it was demonstrated that prototypes are not invariant, but change dependent on context. Similarly, another study by Barsalou, this time Barsalou and Sewell (1984, reported in Barsalou, 1985) found that goodness-of-exemplar structure can change dramatically when people take different perspectives on a concept. Subjects rated exemplars (e.g., *robin, ostrich, swan*) for typicality as *birds*, but assigned them different degrees of centrality according to the cultural perspective (American, African, or Chinese) they had been asked to assume. These results were found for both taxonomic and goal-derived categories. Thus one can flexibly generate different concepts in different contexts.

Again, if we assume incorrectly that prototypes are directly mentally represented, then a third type of problem that arises is that of concept combination, as was demonstrated by Osherson and Smith (1981). To compose a complex concept, exemplars from the constituent prototypes would have to be combined (Lyon & Chater, 1990). For example, a representation of the complex concept *pet fish* would be composed by the combination of a prototypical exemplar for *pet* and a prototypical exemplar for *fish*. However, if one does this, it is unlikely that the combination of *dog* and *cod* will produce the intended meaning of *pet fish*. A *guppy* is a good example of a *pet fish*, but it is not prototypical of either *pets* or *fish*.

Despite the overwhelming evidence against the direct lexical and mental representation of prototypes, the majority of lexical semantic accounts in cognitive linguistics (and accounts in the concept literature in general) employ representations explicitly embodying prototypical structure. Ironically, this is also true in Lakoff's (1987) case study of *over*, despite protestations to the contrary earlier in the same volume. Clearly, the claim that prototypes are constituents of lexical categories is an erroneous one. Prototypes cannot be directly mentally represented, but instead should be viewed as processing effects that any theory should take into account without the need to represent them directly.

Given that Lakoff's (1987) first claim falls through, we can take the second claim. The claim is that lexical semantic theories should embody polysemy directly in the lexicon, rather than treating senses like homonyms. This is a rather curious claim as it is unclear what advantages polysemous representations bestow in terms of case accountability and theoretical value *given that* prototypes are not directly mentally represented. As the first claim falls through, then *sequor* the second claim cannot hold. As the evidence suggests that it is unlikely that prototypes are features of lexical concepts, then the reason for representing polysemy directly is no longer there. A worked example of this can be given with a brief examination of the Brugman and Lakoff analysis of *over*.

Brugman (1981, 1988), Lakoff (1987), and Brugman and Lakoff (1988) outlined nearly 100 different kinds of uses of *over*, all of which are directly represented in the lexicon around three prototypical senses. Recognizing so many senses of a word appears at first sight to solve the case accountability problems that minimally specified theories seem to face. Lakoff commented that there appears to be no option but to represent prototypes directly in the lexicon. However, in the account of *over* given by Lakoff and Brugman, and indeed in the fully specified literature in general, no account is given as to how the correct sense is selected from the lexicon in context, and therefore it does not make any difference whether or not the senses that are selected are represented in relation to each other or not. For example, if one considers the sentence "the plane is over the hill," then how does one know that the appropriate sense of *over* here can be either what Brugman and Lakoff called the *above and no contact schema* or the *by way of above schema* but not the *covering schema*. Some method needs to be postulated to select the relevant schema for the context. If one is then having to generate clever selection rules, essentially the pragmatic principles that listing items in the lexicon in a homonymic sense requires, then there are no advantages aside from the direct representation of prototypicality in such an account. We can conclude that the direct representation of polysemy in the lexicon is therefore ill-motivated.

It can be concluded that the criticisms made by Lakoff (1987) of the classical view of categories and concepts can also be lodged against the fully specified alternatives that he, among others, has proposed. On the one hand, minimally specified accounts try to deal with anomalous use through pragmatics, whereas fully specified accounts have to identify a set of principles for sense selection in context, principles that have not been forthcoming. Therefore, in terms of theoretical value, there is in practice not much to choose between them. This still leaves the problem of case accountability, which seems endemic with the types of account thus far considered. One needs some alternative that avoids the problems thus far encountered.

3. SENSE DELINEATION AND LEXICAL
SPECIFICATION

Given the state of affairs just outlined, one can address the issue of how one gets at lexical representation from language use. There are (at least) three levels of representation one must consider when giving an account of lexical specification (depicted in Fig. 15.3). On the right there is the situation in the world, which in the case of spatial language can be viewed as a spatial scene, itself embedded in a wider context (i.e., wider linguistic and spatial contexts). On the left, there is the knowledge we store about language in memory, and in particular where we store lexical contents for words. The third level, a middle level of representation, relates to language use, and is an interface between information stored permanently in memory about words and the world we are interacting with. Most approaches to the lexical semantics of spatial prepositions delineate senses with the assumption that there is a one-to-one mapping between spatial relations in the world and the meaning of a term. Thus one can take all the visual scenes represented in Fig. 15.3, and classify them into different types of spatial relations, which can then be listed as lexical entries for the preposition *in*. For example, Herskovits (1985, 1986) did precisely this, outlining different lexicalized use types for prepositions that map onto

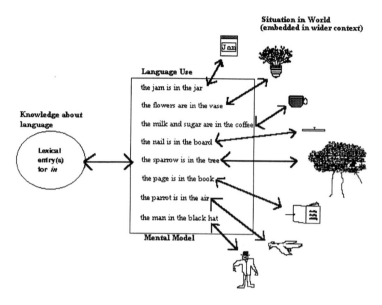

FIG. 15.3. The relationship between lexical entries, language use, and the world.

types of spatial relations in the world (e.g., use types for *in* include "spatial entity in container," "gap/object embedded in physical object," etc.).

Here, there are at least three methodological errors that Herskovits and others make. The first is that the meaning of a spatial expression is not necessarily a direct reflection of the information the preposition brings to the sentence. For example, *the flowers are in the vase* and the *nail is in the board* represent different lexicalized use types for Herskovits (1986), but the difference in meaning of the spatial expressions may be a result of the lexical entries for *flowers, vases, nails,* and *boards* combined with the same lexical entry for *in*. This problem is well known in lexical semantics—one must distinguish between contextual modulation and contextual selection (Cruse, 1986). If one does not recognize this distinction, then one can in principle recognize an infinite number of senses (Bennett, 1975; H. H. Clark, 1983; Johnson-Laird, 1987).

The second error that is made is that researchers confuse categories with lexical concepts. Categorizing the world into different types of spatial relations does not necessarily map onto the lexical entries for spatial prepositions without a principled account of why this should be the case. In fact, there are principled reasons why this is not the case. One must be able to categorize geometric relations in the world before one is able to map these onto language. However, as Crangle and Suppes (1989) stated, "in spite of the spate of articles in the last decade or so on locative expressions, spatial prepositions, and the like, detailed attention to the kinds of geometry needed to give a semantic analysis of the various locative expressions does not seem to have been previously attempted" (p. 399). They went on to argue that a detailed understanding of geometry is required before an adequate characterization of the meaning of spatial language can take place and Suppes (1991) outlined no less than seven different kinds of geometry that may need to be employed to underlie the basic meaning of different spatial prepositions. The striking feature here is the level of complexity required to describe even the most basic of geometric relations, and the variety of types of geometry that exist. If such geometries do underlie spatial language, consonant with the view that spatial language does indeed refer to spatial relations in the world, then one must give an account as to how the correct geometry is selected in context before one can adequately talk about the world. Furthermore, as Suppes commented, there is no reason to believe that a full categorization of the types of geometry that are required can be achieved in the first place. Thus, the assumption that geometry does underlie the semantics of spatial terms appears at best premature given the difficulties in encapsulating the geometry of spatial relations in the world.

Of course, the other objection to sense delineation through categorizing spatial relations is that there is not a one-to-one mapping between spatial relations and lexical entries. If the world were categorizable into distinct spatial geometric relations, then one would expect language to map onto these geometric relations monotonically, but they do not. For example, if an object is not in its canonical orientation the language used to talk about the same geometric relations changes (as illustrated in Figs. 15.1 and 15.2).

To conclude this section, we have seen that there are several key objections to the methodologies used to delineate senses in the spatial preposition literature. The starting point for such an endeavor should be the relationship between language and the world. All one can endeavor to do is to look systematically at the relationship between language use and the world, and to make transcendental inferences regarding lexical entries.

4. SPATIAL PREPOSITIONS AND FUNCTIONAL RELATIONS

A different approach to the semantics of spatial prepositions begins with the basic question of the function of spatial language from the point of view of the organisms that use it (in much the same way as Marr, 1982, started with the issue of what the visual system is for—explanation at the computational level). Garrod and Sanford[3] (1989) and Vandeloise (1984, 1991) argued that functional relations underlie the meaning of the spatial prepositions *in*, *on*, and *at*, together with the notion that mental models are interfaces between language and the spatial world. Functional relations have to do with how objects are interacting with each other, and what the functions of objects are. It is this type of information that is pertinent to an understanding of the spatial world. Garrod and Sanford provided the starting point for functional contents associated with spatial prepositions, and this line was developed by Coventry (1992, 1995). Coventry proposed that the lexical entry for *in* is something like: **in**: functional containment—*in* is appropriate if the ground is conceived of as fulfilling its containment function. Whether or not *in* is appropriate depends on the mental model adopted in the specific situation, where a mental model is defined as a temporary structure in working memory that serves as an interface between language and the spatial world. Such models are built both from the lexical entry and general world knowledge relevant to the situation and from the situation itself. If one takes the encoding problem, there is a spatial scene in the world from which a variety of models can be derived. The appropriateness or otherwise of the spatial preposition is then dependent on the mental model adopted. With the decoding problem, the language itself can suggest a particular model of the scene. The mental model constructed can then be imposed on the actual scene in the world, at which point the visual scene in the world can be said to be a valid, or invalid, visual representation of the language used.

This analysis implicates the importance of a number of factors that determine, for example in the case of *in*, whether the container is fulfilling its function. If it can be demonstrated through movement over time that the container is constraining the location of the figure, then *in* should be appropriate. This can be done through movement of the container where the figure remains in the same position relative to the container, or where the container is sealed, thus blocking movement of the figure beyond the rim of the container, allowing constraint of location over time.

Prior knowledge about objects and how they interact is also a factor that influences language use. For instance calling the same object a *dish* versus a *plate* may suggest different object-specific properties and thus influence prepositional usage. As Michotte (1963) suggested, "It is by coming to know what things do that we learn what they are. What they are for is much more than their shape, their size, and their colour; it is above all what they are capable of doing, or what can be done with them" (p. 3).

---

[3]Talmy (1988) provided a related account involving force dynamics, but his approach still assumes the primacy of spatial relations and thus cannot account for the data presented. See Coventry (1995) for a discussion.

Evidence for this functional analysis comes in the recent empirical work of Coventry and others (Coventry, 1992, 1995; Coventry, Carmichael, & Garrod, 1994; Coventry & Mather, in press; Ferrier, 1991). Coventry and Ferrier initially tested for the influence of functional relations on spatial prepositions, mainly dealing with the preposition *in*. In the case of Coventry's study, subjects were presented with spatial scenes, and were asked to fill in sentences presented with the scenes in order to describe the relationship between the figure and the ground in each scene. As this experiment was presented under the guise of a memory experiment, the sentence completions were spontaneous and natural, and on analysis several key patterns of findings were revealed. Contiguity of movement of figure with ground was found to increase the use of *in*. For example, *in* was used significantly more in Fig. 15.4g as compared with Fig. 15.4h. Thus a demonstration that the container is fulfilling its function by controlling the location of the figure over time clearly increases the use of *in*. Conversely, movement of the figure independently of the ground was found to reduce the use of *in*, although not significantly in all cases (e.g., *in* was used significantly more in Fig. 15.4e than in Fig. 15.4f). Similarly, tilting the container reduces the use of *in* if it looks like the figure may fall out.

In the absence of clear evidence for functional control, other information in the scene may be used to build a model of that scene. Comparing static scenes where geometric relations remained constant, it was found that contact of the figure via other objects (continuity preserved or otherwise) increased the use of *in*. For example, *in* was used more in Fig. 15.4a than in Fig. 15.4b. The use of *in* is therefore clearly underdetermined by geometric relations alone. Additionally, noncontinuity of figure with other objects produced a significant reduction in the use of *in* when the pile was high in static scenes (e.g., *in* was used significantly more in Fig. 15.4c than in Fig. 15.4d).

Effects were also found when comparing static scenes involving a jug and a bowl. *In* was used significantly more with the bowl as ground than with the jug as ground when the pile was high (e.g., *in* was used significantly more in Fig. 15.4i than Fig. 15.4j). Investigating this apparent object-specific function effect further, Coventry et al. (1994) found that adding liquid to a jug further decreases the use (and ratings) of *in*, but makes no difference in the case of the bowl. Thus the addition of water appears to make the object-specific function of the jug (to contain liquids) more salient, further reducing the appropriateness of the container as a container of solids.

Functionality effects have been extended by Coventry et al. (1994) to the prepositions *on*, *over*, and *beside*, and Coventry and Mather (in press) provided an account of the spatial uses of *over* that relies heavily on object

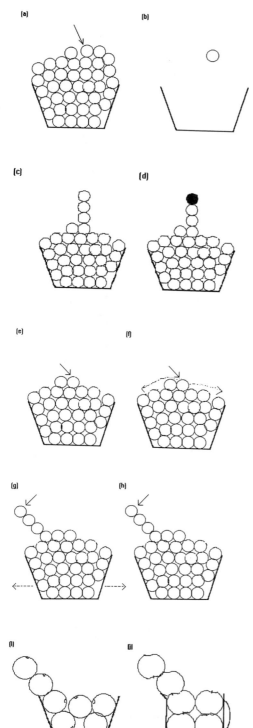

FIG. 15.4. Examples of scene contrasts used in the studies by Coventry et al.

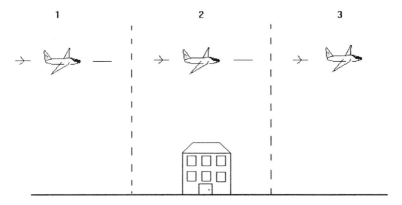

FIG. 15.5. Scene used by Coventry and Mather (in press).

knowledge. Furthermore, Coventry and Mather demonstrated that a semantics for spatial expressions is dependent on general world knowledge outside the lexicon. They proposed that *over* differs from *above*[4] in that *over* is extremely sensitive to object-specific knowledge and functional relations between objects that extends to knowledge of (naive) physics. Subjects were presented with a diagram of a plane and a building segmented into three sections (illustrated in Fig. 15.5 above). They were then asked where the plane should be for the expression *the plane is over the building* to hold either without a context, or in the context of the plane on a bombing mission to bomb the building. The use and ratings for the segments were significantly different between conditions with the biggest difference occuring in the use and rating of the first segment. Furthermore, it was found that both the ratings and use of the expression correlated significantly in the case of the bomb scenario with judgments of where subjects thought the bomb should be dropped to successfully hit the building. Thus an understanding of a spatial expression is reliant on information outside the language for its instantiation.

Other types of evidence for the importance of functional relations come from developmental research. The order of acquisition of spatial prepositions in English (and cross-linguistically; see Sinha, Thorseng, Hayashi, & Plunkett, 1994) supports the argument that terms embodying functional relations are acquired earlier than purely spatial terms. For instance, *over* is acquired before *above*, although this would not be predicted by any criteria other than functionality (e.g., frequency of input or complexity). E. V. Clark (1977) and Bower (1974) also reported evidence suggesting the early understanding of functionality in young children.

---

[4]Spatial prepositions like *above* can be regarded as minority spatial terms as they are not influenced by the kinds of functionality effects outlined here.

5. CONCLUSION: SOME MORALS
FOR AI INTERFACES

The approach and evidence reviewed in the preceding sections, together with other evidence not reviewed here (see Coventry, 1992, 1995; Coventry et al., 1994), illustrates that attempts to interface spatial language and the spatial world that rely on specifying geometric relations between figure and ground in visual scenes are unlikely to be successful. Functional relations clearly influence spatial language use, and furthermore involve information from general world knowledge about Newtonian laws as well as information regarding object functions that may reside in the lexicon.

Before the decoding and encoding problems can be solved, one needs to construct an intermediate representation (mental model) between the input (e.g., a spatial scene) and information stored in memory (e.g., lexical semantics, general world knowledge). Whether or not a particular spatial term is appropriate in the case of most spatial prepositions can be understood only with reference to the model. The construction of models is influenced by a myriad of factors, including continuity/discontinuity of figure with ground, the specific functions of objects, relative of movement of figure and ground over time, and other types of knowledge regarding expected interactions. Thus the empirical data regarding language understanding and comprehension point clearly to factors that need to be incorporated into the design of AI systems to improve cognitive plausibility and thus facilitate a solution to the decoding and encoding problems.

REFERENCES

Armstrong, S. L., Gleitman, L., & Gleitman, H. (1983). What some concepts might not be. *Cognition, 13,* 263–308.

Barsalou, L. W. (1983). Ad-hoc categories. *Memory & Cognition, 11,* 211–227.

Barsalou, L. W. (1985). Ideals, central tendency and frequency of instantiation as determiners of graded structure in categories. *Journal of Experimental Psychology: Learning, Memory and Cognition, 11,* 629–654.

Bennett, D. C. (1975). *Spatial and temporal uses of English prepositions: An essay in stratificational semantics.* London: Longman.

Bower, T. (1974). *Development in infancy.* San Francisco: Freeman.

Brugman, C. (1981). *The story of "over."* Unpublished master's thesis, University of California, Berkeley.

Brugman, C. (1988). *The story of "over": Polysemy, semantics and the structure of the lexicon.* Garland Press.

Brugman, C., & Lakoff, G. (1988). Cognitive topology and lexical networks. In G. W. Cottrell, S. Small, & M. K. Tannenhause (Eds.), *Lexical ambiguity resolution: Perspectives from psycholinguistics, neuropsychology and artificial intelligence.* San Mateo, CA: Morgan Kaufman.

Casad, E. (1982). *Cora locations and structured imagery*. Unpublished doctoral dissertation, University of California, San Diego.

Clark, E. V. (1977). Strategies and the mapping problem in first language acquisition. In T. MacNamara (Ed.), *Language learning and thought*. New York: Academic Press.

Clark, H. H. (1983). Making sense of Nonce Sense. In G. B. F. d'Arcais & R. J. Jarvella (Eds.), *The process of language understanding* (pp. 297–331). Chichester, England: Wiley.

Cooper, G. S. (1968). *A semantic analysis of English locative prepositions* (Report No. 1587). Bolt, Beranek & Newman.

Coventry, K. R. (1992). *Spatial prepositions and functional relations: The case for minimally specified lexical entries*. Unpublished doctoral dissertation, University of Edinburgh, Edinburgh, Scotland.

Coventry, K. R. (1995). *Spatial prepositions, functional relations and lexical specification*. Working paper, Department of Psychology, University of Plymouth, Plymouth, England.

Coventry, K. R., Carmichael, R., & Garrod, S. C. (1994). Spatial prepositions, object-specific function and task requirements. *Journal of Semantics, 11*, 289–309.

Coventry, K. R., & Garrod, S. C. (in preparation). *Seeing, saying and acting. The psychological semantics of spatial prepositions* (Essays in Cognitive Psychology Series). Mahwah, NJ: Lawrence Erlbaum Associates.

Coventry, K. R., & Mather, G. (in press). The real story of *Over*. In P. Olivier & W. Maass (Eds.), *Representations between vision and language*. New York: Springer-Verlag.

Crangle, C., & Suppes, P. (1989). Geometric semantics for spatial prepositions. *Midwest Studies in Philosophy, XIX*, 399–422.

Cruse, D. A. (1986). *Lexical semantics*. Cambridge, England: Cambridge University Press.

Ferrier, G. (1991). *An experimental study to investigate the factors that contribute to a functional geometry of spatial prepositions*. Maxi project, Department of Psychology, University of Glasgow, Glasgow, Scotland.

Garrod, S. C., & Sanford, A. J. (1977). Interpreting anaphoric relations: The integration of semantic information while reading. *Journal of Verbal Learning and Verbal Behavior, 16*, 77–90.

Garrod, S. C., & Sanford, A. J. (1989). Discourse models as interfaces between language and the spatial world. *Journal of Semantics, 6*, 147–160.

Hawkins, B. W. (1984). *The semantics of English spatial prepositions*. Unpublished doctoral dissertation, University of California, San Diego.

Herskovits, A. (1985). Semantics and pragmatics of spatial prepositions. *Cognitive Science, 9*, 341–378.

Herskovits, A. (1986). *Language and spatial cognition. An interdisciplinary study of the prepositions on English*. Cambridge, England: Cambridge University Press.

Jakobson, R. (1932). 'Zur Struktur des russischen Verbums,' *Charisteria V, Mathesio Oblata* (pp. 74–83). Prague: Cercle Linguistique de Prague. (Reprinted in Hamp, Householder and Austerlitz, 1966, pp. 22–30).

Jakobson, R. (1936). 'Beitag zur allgemeinen Kasuslehre: Gesamtbedeutungen der russischen Kasus' (pp. 240–288). (Reprinted in Hamp, Householder and Austerlitz, 1966, pp. 51–89).

Janda, L. (1984). *A semantic analysis of Russian verbal prefixes ZA-, PERE-, DO-, and OT-*. Unpublished doctoral dissertation, University of California, Los Angeles.

Johnson-Laird, P. N. (1987). The mental representation of the meaning of words. *Cognition, 25*, 189–211.

Lakoff, G. (1987). *Women, fire and dangerous things*. Chicago: University of Chicago Press.

Leech, G. N. (1969). *Towards a semantic description of English*. London: Longman.

Lindkvist, K. G. (1950). Studies on the local sense of the prepositions IN, AT, ON, and TO in modern English. In *Lund Series in English* (Vol. 22). Lund and Copenhagen: Munksgaard.

Lindner, S. (1981). *A lexico-semantic analysis of verb-particle constructions with up and out.* Unpublished doctoral dissertation, University of California, San Diego.

Lyon, J., & Chater, N. (1990). Localist and globalist approaches to concepts. In K. J. Gilhooly (Ed.), *Lines of thinking* (Vol. 1). Chichester, England: Wiley.

Marr, D. (1982). *Vision. A computational investigation into the human representation of visual information.* New York: W. H. Freeman.

Michotte, A. (1963). *The perception of causality.* London: Methuen.

Miller, G. A., & Johnson-Laird, P. N. (1976). *Language and perception.* Cambridge, MA: Harvard University Press.

Osherson, D. N., & Smith, E. E. (1981). On the adequacy of prototype theory as a theory of concepts. *Cognition, 9,* 35–58.

Retz-Schmidt, G. (1988, Summer). Various views on spatial prepositions. *AI Magazine,* pp. 95–105.

Rosch, E. (1973). Natural categories. *Cognitive Psychology, 4,* 328–350.

Rosch, E. (1975). Cognitive representations of semantic categories. *Journal of Experimental Psychology: General, 104,* 192–233.

Rosch, E. (1977). Human categorisation. In N. Warren (Ed.), *Advances in cross-cultural psychology* (Vol. 7). London: Academic Press.

Rosch, E., & Mervis, C. B. (1975). Family resemblances. *Cognitive Psychology, 7,* 573–605.

Roth, E. M., & Shoben, E. J. (1983). The effect of context on the structure of categories. *Cognitive Psychology, 15,* 346–78.

Rudzka-Ostyn, B. (1983). *Cognitive grammar and the structure of Dutch uit and Polish wy.* Trier, Germany: Linguistic Agency, University of Trier.

Sandhagen, H. (1956). *Studies on the temporal senses of the prepositions AT, ON, IN, BY and FOR in present-day English.* Trelleborg, Sweden: Author.

Sinha, C., Thorseng, L. A., Hayashi, M., & Plunkett, K. (1994). Comparative spatial semantics and language acquisition: Evidence from Danish, English and Japanese. *Journal of Semantics, 11,* 261–288.

Suppes, P. (1991). The principle of invariance with special reference with special reference to perception. In J. Doignon & J. Falmagne (Eds.), *Mathematical psychology: Current developments* (pp. 35–53). New York: Springer-Verlag.

Talmy, L. (1988). Force dynamics in language and cognition. *Cognitive Science, 12,* 49–100.

Vandeloise, C. (1984). *The description of space in French.* Unpublished doctoral dissertation, University of California, San Diego.

Vandeloise, C. (1991). *Spatial prepositions. A case study from French.* Chicago: University of Chicago Press.

16

The Representation of Space and Spatial Language: Challenges for Cognitive Science

Barbara Landau
University of Delaware

Edward Munnich
MIT

Among the most impressive of human capacities is the ease with which we express aspects of our spatial experience. Even from the earliest years, it is a trivial task for us to talk about where we are, where we have been, the objects we have encountered and what we have done with them, our location when we did so, and how things turned out. Despite the apparent ease with which we do this, scientists who have probed recently into the nature of the representations underlying this capacity have been impressed with how complex and subtle our knowledge must be in order to connect language with what we perceive and know about the spatial world. For psychologists, understanding the nature of these fundamental representations is, of course, a goal in and of itself; for computer scientists, understanding how human language maps onto spatial representations is a prerequisite for creating computer systems that can receive natural language instructions and carry out natural human actions.

Given these two related goals, it is apparent that there is widespread agreement on two things: First, there is a critical need to understand how humans represent objects, actions, forces, and locations and how language maps onto these. And second, the task of doing so is bound to be both difficult and complex. In these remarks, we would like to highlight a few areas in which scientists have just begun to touch on the subtlety of spatial language and its relationship to human spatial representations.

1. THE PROBLEM OF AMBIGUITY:
PARTITIONING THE SCENE

First we consider some problems and issues raised by thinking about how language maps onto space; in particular, the role that language itself plays in guiding our interpretation of spatial scenes. The basic problem here is that any single scene is multiply ambiguous with respect to our interpretations: Static scenes include many objects and many possible spatial relationships; scenes with motion additionally include motions, paths, and therefore many changing spatial relationships. Given the natural units of language and also given that spatial scenes (the objects, the locations, the motions, the paths) will be partitioned somewhat differently in different languages, the question arises which units we should consider central to our investigation. That is, which terms or combinations of terms express which spatial meanings?

A sensible way to start would be to consider the units that, in a given language, are the canonical carriers of spatial meaning. Many investigators have proceeded this way; for example, in a number of articles, investigators focus on spatial prepositions such as *on, off, above, below*, and the phrases in which they are found, for example, "on the mat," "off the table," "above the building." These expressions certainly capture a number of basic spatial relations.

However, even in English (where places are canonically encoded as PPs; see Jackendoff, 1983), the complete expression of the spatial relationship is bound to be much more complex: The lexical expression of spatial relationships spills well into other form classes. Furthermore, some elements of spatial meaning are derived from the complete syntactic contexts in which these elements occur.

Let's start by considering form class. In English, the prepositions *on, against, off,* and so forth, express attachment relationships. Even the preposition *on* can be used for a diverse set of relationships, for example: The cup is *on* the table/ The handle is *on* the cup/ The decal is *on* the cup/ The cup is *on* the hook. The first use of *on* encodes support, the second and third encode attachment, and the fourth encodes support by hanging.

However, English also possesses many verbs of attachment that can also encode these and other attachment relationships: *attach, suspend, hang, stick, cling, adhere, separate, disconnect,* and *disentangle,* to name a few. Notice that three of the relationships encoded by *on* can be expressed readily using different English verbs. (Actually, the verbs of attachment form a potentially infinite class, given the productivity of denominal verbs in English, e.g., *anchor, band, belt, bolt, bracket, buckle, button;* see Levin, 1993). Many of these express subtle variations within force-dynamic relationships. For example, whereas *"attaching* X to Y" can involve any of a large class of specific methods and final relationships, *sticking* or *gluing* involve more specific means of attachment; *"suspending* (or hanging) X *from* Y" involves attachment by a point, rather than a surface.

Further, different languages capture somewhat different classes of relationships with spatial prepositions. For example, Dutch possesses terms that distinguish between attachment by a point versus a surface—both covered by English *on*; Spanish possesses a single term *en* covering English *on* and *in*. Moreover, although English does capture many basic spatial relationships using prepositions, languages such as Korean or Mandarin appear to favor verbs as principal carriers of spatial information (although these languages also have spatial adpositions, they appear to be somewhat more limited than those in English). Clearly, in order to completely capture how language represents space—and especially, to use language to disambiguate spatial scenes—it will be important to understand how different languages use different form class elements to partition spatial scenes. It goes without saying that this will also be critical in the study of how to effect interlingual translation (as outlined by Voss, Dorr, & Sencan, chap. 8 of this volume).

But even broadening the domain to other predicates is not enough. Investigators have long recognized that an important aspect of spatial meaning is carried by the sentential frame in which the preposition occurs. One example concerns the so-called asymmetry of spatial prepositions. As Talmy and others have noted, the preposition *near* is logically symmetrical: If X is near Y, then Y must be near X. However, when we insert *near* into a sentence, the compositional meaning expresses an asymmetrical relationship: It is perfectly natural to say "The bicycle is near the house" but slightly odd to say "The house is near the bicycle." In fact, the latter sentence elicits an image of a tiny (perhaps toy) house near a larger (perhaps real) bicycle. In general, what Talmy called the "figure" (the thing being located) is likely to be small and mobile (or potentially mobile), whereas the "ground" (the reference object) is likely to be larger and stable (or, at least, less mobile)—that is, it has characteristics that make it a good reference object.

Similar asymmetries appear for other spatial predicates; for example, compare "The shutter is attached to the house" to "The house is attached to the shutter" or "The handle was glued to the cup" versus "The cup was glued to the handle." In each case, there is a more natural reading based on the usual roles of cups to handles and houses to shutters.

The broad phenomenon of asymmetry was recently studied by L. Gleitman and colleagues, who have shown asymmetries in people's judgments of nonspatial (though logically symmetrical) predicates (Gleitman, Gleitman, Miller, & Ostrin, 1996). For example, compare (a) to (b), taken from Gleitman et al:

(a) My husband met the Pope. (b) The Pope met my husband.
OR (a) My sister met Meryl Streep. (b) Meryl Streep met my sister.

In each case, the individual considered more important (perhaps, the more natural reference object in the social domain) is most naturally located in the object position. Note also that the second example of each set induces a somewhat different construal: For "The house is attached to the shutter" or "The cup is glued to the handle" we assume some scenario in which the house or the cup is somehow brought to the location of the shutter or handle.

What is happening here is that properties of the sentential frame itself—specifically, the positions in which the noun phrases (NPs) occur relative to the predicate—have the effect of coercing the objects into a certain dominance relationship: The noun phrase in subject position plays the role of "figure" in the speaker's mental representation of the event, whereas the noun phrase in object position plays the role of "reference object." Each of these can play different traditional thematic roles—agent or theme in subject position, object or theme in direct object position, and so on. But what they have in common is their relative positions in what Fisher (in press) called the "conceptual hierarchy." The *syntactic* positions of the two noun phrases dictate their asymmetry. This is why "the house" in "The house is near the bicycle" plays the role of figure object, inducing a construal of a small, mobile house—probably a toy.

Given this, the role of syntactic structure in guiding the mapping between spatial language and spatial representations could play an important role in establishing the basic relationship of figure and reference object. To our knowledge, most computer applications assume that the user knows which is the figure and which is the reference object. However, it is not so self-evident that this can be determined by observation of a scene alone—that is, without linguistic input. For any spatial relationship, there are minimally two objects, hence at least two possibilities: A is above B/ B is below A; A is on top of B/ B is under A; a coat is hung on a hook/ the hook holds the coat, and so forth. By providing a user with only a labeled obect pair (e.g., coat, hook) there are multiple possible spatial relationships, each of which could felicitously describe the scene. In contrast, with a full sentence, one specifies both the relative roles of the two objects—which is figure and which is reference object—and the spatial relationship. Having told a user (or the computer) that "X is on Y" or "X is near Y," and so on, the user with knowledge of syntactic structure (hence the relative dominance of the NP arguments) can directly determine which is the "figure" and which is the "reference object." Knowing which is the reference object then allows the listener/user to set the reference frame on this object.

This type of scheme implies some sort of interaction between linguistic input and the spatial representations they describe. It suggests the possibility that people may use such syntactic information as a guide to exploring arrays and locating objects. As shown by Spivey-Knowlton, Tanenhaus, Eberhard, and Sedivy (chap. 12 of this volume), such interactions between syntactic structures and spatial representations are likely during online processing, suggesting that information from syntactic structure can serve as a powerful guide to disambiguating spatial contexts, and vice versa.

2. MORE PROBLEMS OF AMBIGUITY: ESTABLISHING FRAMES OF REFERENCE

Once figure and reference object have been identified, and we have isolated the elements key to expressing spatial location, a number of other problems surface. One issue addressed by a number of investigators is how we can determine which frame or frames of reference might be most natural for establishing location. Most investigators have agreed that there are multiple frames of reference available to humans speakers and listeners; and a significant problem for communicating about location is to determine which frame is being used at what time. For example, if one says "The ball is above the chair," a number of possibilities arise—to name just two possibilities, the ball might be located along the principal axis (and its extension) of the chair, or along the environmentally vertical axis (see Levelt, 1996, for discussion). This problem of ambiguity is both compounded and mitigated by our use of reference frames. It is compounded because (as Schober, chap. 14 of this volume, has shown), speakers can switch among reference frames quite flexibly, and listeners do follow. And it is mitigated by the fact that humans do appear to have biases in which reference systems they will use on which occasions (Carlson-Radvansky & Irwin, 1993). As Gordon Logan (1996) showed, flexible use of different reference frames appears to be a fundamental property of human attentional systems. That is, we are capable of using multiple reference frames and it is natural for us to switch flexibly among them.

Actually, things are probably even more complex than the investigators in this volume have outlined. Whereas most investigators accept at least three relevant reference systems (object centered, ego centered, environment centered), Jackendoff (1996) recently argued that there are actually at least *eight* different reference systems that can be used to ascertain the axes (and therefore the relevant spatial regions) of an object, that is, its top, bottom, front, back, and those regions that are above, below, in front of, behind, and so on. He described four "intrinsic" frames and four "environmental" frames.

Within the intrinsic frames, Jackendoff (1996) listed the geometric frame (using the actual geometry of the object to determine its axes), the motion frame (using the direction of motion to establish the front-back axis), and two other frames that depend on functional properties of the object. The "canonical orientation" frame designates the top/bottom of an object in its normal orientation (despite its current orientation), and the "canonical encounter" frame designates various parts by normal interactions (e.g., the "front" of the house is where one enters; the "top" of a camera is the part where the on-off button might be found).

The environmental frames provide four more ways of determining the axes of an object. These include the gravitational frame (determined by gravity, irrespective of object orientation), the geographical frame (NEWS), the contextual frame (imposed by some surrounding object such as the page on which an object is drawn), and the observer frame (e.g., establishing the "front" as the side facing the observer, in English).

In English, the language itself does not give much of a hint about which frame of reference might apply (and this may well be true of most other languages). There are several exceptions, however. One is the use of the NEWS system, which uses distinct terms for the geographical reference system. (This appears to be the case for Tzeltal as well; see Levinson, 1996). A second is the use of the determiner *the* in the noun phrase "the X," where X is the spatial region. For example, the region that is "on top of X" could apply to any of the frames described. But the region that is "on *the* top of X" could only apply to intrinsic frames of reference. And a third is the differential use of axial terms such as *length* and *height* (subjects tend to link the former to the object's principal axis, whereas the latter is linked to the environmental vertical; see Narasimhan, described by Jackendoff, 1996). Other than these, however, how do we know which frame of reference is being used, and/or which to use? This is obviously critical to the task of communication.

As a number of contributors point out, the frame of reference that is most natural for a given scene may depend on a variety of factors including the type of layout (a city vs. a tabletop), the kind of object being considered (e.g., a camera, whose principal axis is environmentally horizontal as we hold it, *vs.* a cellular telephone, whose principal axis is environmentally vertical as we hold it), and speaker/hearer use, among other things. The case of specific object knowledge would seem to be a particularly thorny problem for theories of spatial representations, as well as for computer schemes.

Consider the camera/phone example: Both objects have roughly the same three-dimensional (3-D) shape (a rectilinear volume), but their functions, hence canonical usage, determine which is the principal axis. If specific knowledge of object functions affects how the reference system is set up, then we must ask whether we want to *list* a potentially infinite number of objects and functions directly in the spatial representation (a solution that would seem somewhat cumbersome) or refer this spatial representation to ancillary storage containing object-specific knowledge. In any case, the fact that the nature of the object matters a great deal in deciding on its principal axes indicates that stored knowledge of objects will likely play an important role in locating spatial regions.

But what about *novel* objects? Although characterizing the shapes of novel objects has been a focus of object recognition studies, it appears to be less well studied in the context of spatial terms. How does a speaker/hearer decide on what is the top, bottom, front, or back of a novel object? Here, we cannot appeal to specific knowledge. What else is available?

One particularly rich source of information may be the perceptual structure of the object itself. For example, an object's principal axis of symmetry will usually (though not always) specify the top/bottom axis of the object. In addition, as Narasimhan (in Jackendoff, 1996) showed, subjects will often assume that a flattened part of a (novel) object is its base, therefore its bottom. Such use of perceptual structure can be important in ascertaining the location of intrinsic object axes for novel objects— something that is critical for object recognition as well. There may well be additional perceptual clues to the "fronts" of objects (in artifacts, often the most complex) and the "tops" (ditto), but these are hardly foolproof. Clearly, both normal functional interactions with objects *and* the flexibility of humans in assigning and reassigning reference frames will be critical in fully understanding how we set up an object's reference frame, and how we use it with other reference frames to locate things.

In summary, the issue of reference systems is complex, and its full treatment will likely require that we consider sources of information such as (a) linguistic structure, (b) perceptual properties, (c) functional properties, (d) speaker's intentions and listener's biases.

3. BEYOND GEOMETRY: THE IMPORTANT ROLE OF FORCE-DYNAMICS

The last topic we consider is our current favorite, both because it seems so critical to fully understanding spatial language, and also because it appears to tap into rich knowledge systems that might arise early in human development. The topic is the role of force-dynamics in the representation of spatial terms.

A number of investigators have recently argued that geometric representations are either inadequate or fully unsuited to describing the basic meanings of spatial terms (e.g., Vandeloise, 1991; see also Coventry, chap. 15 of this volume). Coventry has suggested that understanding even basic spatial prepositions such as *in* requires that one consider both force-dynamic properties of objects—how they interact with each other—as well as specific functions carried by particular objects. Although we think that Coventry is on the right track suggesting the importance of these factors, we would like to suggest that, even with the clear importance of these factors in adult usage, the issues remain quite complex.

First, a cautionary comment: Although force-dynamics appears to be important for a variety of terms, it does not replace the geometric approach, but rather, adds a necessary element. For example, Coventry reports on a number of experiments showing modulation of adults' usage of *in*, depending on the force-dynamic (or interactive) properties of the figure object relative to the reference object as well as intermediate objects. However, what these force-dynamic properties appear to do is to *define* the geometric area within which the figure can be considered to be "in" the reference object. So, in his example of a moving reference object (compared to a static one), the motion of the reference object has the effect of expanding the geometrical region that is acceptable. A question well worth pursuing is the nature of human force-dynamic knowledge of objects, and how these enter into the earliest acquisition of terms such as *in/on*, which surely have a force-dynamic component.

Consider another example, using terms that have been somewhat less likely candidates for force-dynamic treatment: These are axial terms such as *over/above*. In a strict geometrical treatment, the region of relevance is defined by (let's say) a vertical axis centered on the reference object. As shown by Hayward and Tarr (1995) for English-speaking adults, and in our lab for 3-year-olds, the fully specified region of acceptability almost never subtends just that axis and its extension. Rather, the acceptable region forms a *V*-shaped wedge leading out roughly from the center of the reference object upward. This finding has recently been replicated among native Japanese speakers (Munnich, 1997). This indicates that, for stable objects, there is a well-defined geometric region for axial terms.

Do force-dynamic properties modulate these geometric representations? It seems likely that they do. For example, compare the region that might be acceptable for a static decorative piece anchored "over" the building to any flying object that moves "over" the building. Or, compare the region that would be acceptable for a mouse running "in front of" a mouse hole, compared to the probably broader and ever-changing region defined for that mouse running "in front of" the pursuing cat.

Aside from force-dynamics, Coventry has also shown that subjects' judgment of location for *over* differs when their judgment concerns a plane merely flying over a building compared to flying on a mission to bomb the building; he concludes from this that considerable world knowledge is required to specify the meaning of *over*. To modify this a bit, we would argue that there still exists a geometrical region for *over* that is well specified (and depends on the axes projecting from the reference object), but that it can be modified by force-dynamics. Some force-dynamic properties may be quite general (e.g., objects fall if not supported), whereas others may require a good deal of object-specific knowledge.

For psychologists, this issue is more than terminology or notation. Rather, the question becomes: Do humans possess some fundamental principles that determine the structure of regions; and do these principles interact with idiosyncratic information that must be learned? For someone interested in acquisition of spatial terms, it is striking that early acquisition appears to be relatively error-free (as far as it has been investigated); and this raises the important question of whether force-dynamics and/or specific object knowledge comes to affect the representation of spatial terms later in development than geometric knowledge, or whether all three factors interact from the beginning. We assume that the correct theories of these will have implications for how computer implementations package up the different components, and how they are designed to interact with each other.

In conclusion, we would like to emphasize that our current understanding of how humans map together space and language is far from complete. However, we consider it a sign of great health that so many important questions can be raised by thinking about the fundamental problem of how we talk about space.

ACKNOWLEDGMENTS

This chapter was prepared as an invited commentary to the Workshop on Representation and Processing of Spatial Expressions, International Joint Conference on Artificial Intelligence, Montreal, Canada, August 1995.

REFERENCES

Carlson-Radvansky, L. A., & Irwin, D. (1993). Frames of reference in vision and language: Where is above? *Cognition, 46*(3), 223–244.

Fisher, C. (in press). From form to meaning: A role for structural analogy in the acquisition of language. In H. W. Reese (Ed.), *Advances in child development and behavior*. New York: Academic Press.

Gleitman, L., Gleitman, L., Miller, C., & Ostrin, R. (1996). Similar, and similar concepts. *Cognition, 58*(3), 321–376.

Hayward, W. G., & Tarr, M. (1995). Spatial language and spatial representation. *Cognition, 55,* 39–84.

Jackendoff, R. (1983). *Semantics and cognition.* Cambridge, MA: MIT Press.

Jackendoff, R. (1996). The architecture of the linguistic-spatial interface. In P. Bloom, M. Peterson, L. Nadel, & M. Garrett (Eds.), *Language and space* (pp. 1–30). Cambridge, MA: MIT Press.

Levelt, W. J. M. (1996). Perspective-taking in spatial descriptions. In P. Bloom, M. Peterson, L. Nadel, & M. Garrett (Eds.), *Language and space* (pp. 77–108). Cambridge, MA: MIT Press.

Levin, B. (1993). *English verb classes and alternations.* Chicago: University of Chicago Press.

Levinson, S. (1996). In P. Bloom, M. Peterson, L. Nadel, & M. Garrett (Eds.), *Language and space* (pp. 109–170). Cambridge, MA: MIT Press.

Logan, G. D., & Sadler, D. (1996). A computational analysis of the apprehension of spatial relations. In P. Bloom, M. Peterson, L. Nadel, & M. Garrett (Eds.), *Language and space* (pp. 493–530). Cambridge, MA: MIT Press.

Munnich, E. (1997). Axial terms: A cross-linguistic comparison. Master's thesis, University of California, Irvine.

Vandeloise, C. (1991). *Spatial prepositions: A case study from French.* Chicago: University of Chicago Press.

Author Index[1]

---

[1]These are text listings only. There is a corresponding reference for every text listing at the end of each chapter. The reference page numbers are not listed in this author index.

273

Subject Index

Contributors

Anselm Blocher
(anselm@cs.uni-sb.de)
SFB 314, Project Vitra
Department of Computer Science
Universität des Saarlandes
D-66123 Saarbrücken
Germany

David J. Bryant
(bryant@neu.edu)
NYMA, Inc.
500 Scarborough Dr., Suite 205
Egg Harbor Township, NJ 08234

Hilary Buxton
(hilaryb@cogs.susx.ac.uk)
School of Cognitive and
  Computing Sciences
University of Sussex
Falmer
Brighton BN1 9QH
UK

Kenny R. Coventry
(KCoventry@plymouth.ac.uk)
Department of Psychology
Faculty of Human Sciences
University of Plymouth, Drake Circus
Plymouth PL4 8AA
UK

Vittorio Di Tomaso
(ditomaso@alphalinguistica.sns.it)
Laboratorio di Linguistica
Scuola Normale Superiore
Piazza dei Cavalieri 7
56123 Pisa
Italy

Bonnie J. Dorr
Department of Computer Science
University of Maryland
College Park, MD 20742 USA

Kathleen M. Eberhard
Department of Psychology
Notre Dame University
South Bend, IN 46556

Geoffrey Edwards
(edwardsg@vm1.ulaval.ca)
Centre de recherche en géomatique
Université Laval, Ste-Foy
Québec G1K 7P4
Canada

Lidia Fraczak
(fraczak@limsi.fr)
UPS, LIMSI
bat. 508
BP 133
F-91 403 Orsay Cedex
France

Thomas Fuhr
(fuhr@techfak.uni-bielefeld.de)
AG Angewandte Information
Technische Fakultät
Universität Bielefeld
Postfach 100131
33501 Bielefeld
Germany

Klaus-Peter Gapp
FB-14 Informatik IV
Universität des Saarlandes
Postfach 151150
D66041 Saarbrücken
Germany

P. Bryan Heidorn
(pheidorn@uiuc.edu)
Graduate School of Library and
    Information Science
University of Illinois
Library and Information Science Building
501 East Daniel Street
Champaign, IL 61820-6212
USA

Annette Herskovits
Institute of Cognitive Studies
University of California, Berkeley
Barrows Hall, 608
Berkeley, CA 94720

Richard J. Howarth
(richardh@cogs.susx.ac.uk)
School of Cognitive and Computing
    Sciences
University of Sussex
Falmer
Brighton BN1 9QH
UK

Barbara Landau
(blandau@udel.edu)
Department of Psychology
Wolf Hall
University of Delaware
Newark, DE 19716
USA
and
Department of Cognitive Science
University of California
Irvine, CA 92717
USA

Leonardo Lesmo
Centro di Scienze Cognitive
Università di Torino
Via Lagrange 3 - 10123
Torino
Italy

Vincenzo Lombardo
Dipartmento di Informatica
Università di Torino
C.so Svizzerra 185 - 10149
Torino
Italy

Bernard Moulin
(moulin@ift.ulaval.ca)
Centre de recherche en géomatique
Université Laval, Ste-Foy
Québec G1K 7P4
Canada

Amitabha Mukerjee
(amit@iitk.ernet.in)
Department of Mechanical Engineering
    and Center for Robotics
Indian Institute of Technology
Kanpur 208016, UP
India

Edward Munnich
MIT
14N-321
Cambridge, MA 02139

Patrick Olivier
Department of Computer Science
Penglais
Aberystwyth
Dyfed SY23 3DB
Wales, UK

Gerhard Sagerer
AG Angewandte Information
Technische Fakultät
Universität Bielefeld
Postfach 100131
33501 Bielefeld
Germany

Christian Scheering
AG Angewandte Information
Technische Fakultät
Universität Bielefeld
Postfach 100131
33501 Bielefeld
Germany

Michael F. Schober
(schober@newschool.edu)
Psychology Department
Graduate Faculty
New School for Social Research
65 Fifth Avenue
New York, NY 10003
USA

Julie C. Sedivy
Department of Linguistics
University of Rochester
Rochester, NY 14627
USA

Gudrun Socher
AG Angewandte Information
Technische Fakultät
Universität Bielefeld
Postfach 100131
33501 Bielefeld
Germany

Michael J. Spivey-Knowlton
Department of Psychology
236 Uris Hall
Cornell University
Ithaca, NY 14853-7601
USA

Eva Stopp
(eva@cs.uni-sb.de)
SFB 314, Project VITRA
Department of Computer Science
Universität des Saarlandes
D-66123 Saarbrücken
Germany

Michael K. Tanenhaus
Department of Brain and
　Cognitive Sciences
University of Rochester
Rochester, NY 14627
USA

M. Ülkü Şencan
Department of Computer Science
University of Maryland
College Park, MD 20742
USA

Clare R. Voss
(voss@cs.umd.edu)
Department of Computer Science
University of Maryland
College Park, MD 20742
USA